Sekem

Ibrahim Abouleish

Sekem

A Sustainable Community in the Egyptian Desert

Floris Books

Translated by Anna Cardwell

Photographs by Markus Kirchgessner
For details see Photographic Acknowledgments (page 229)

First published in German as *Die Sekem-Vision* in 2004
by Verlag Johannes M. Mayer & Co. GmbH, Stuttgart
First published in English in 2005 by Floris Books, Edinburgh

© 2004 Verlag Johannes M. Mayer & Co. GmbH, Stuttgart, Berlin
This translation © 2005 by Floris Books, Edinburgh

All rights reserved. No part of this publication may
be reproduced without the prior permission
of Floris Books, 15 Harrison Gardens, Edinburgh.
www.florisbooks.co.uk

British Library CIP Data available

ISBN 0-86315-532-4

Produced in Poland by Polskabook

*For all the thousand of people who through
their efforts have contributed to Sekem*

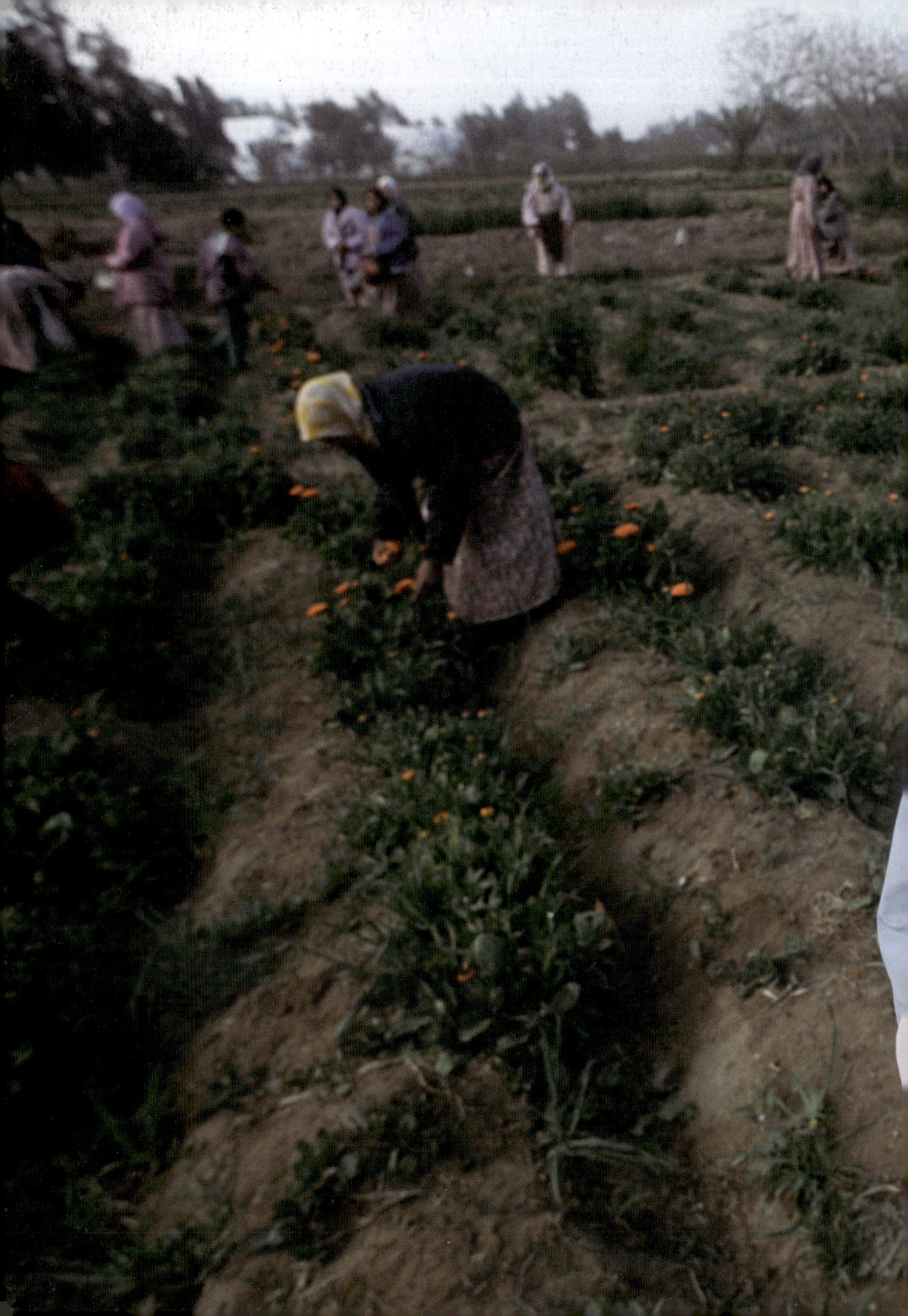

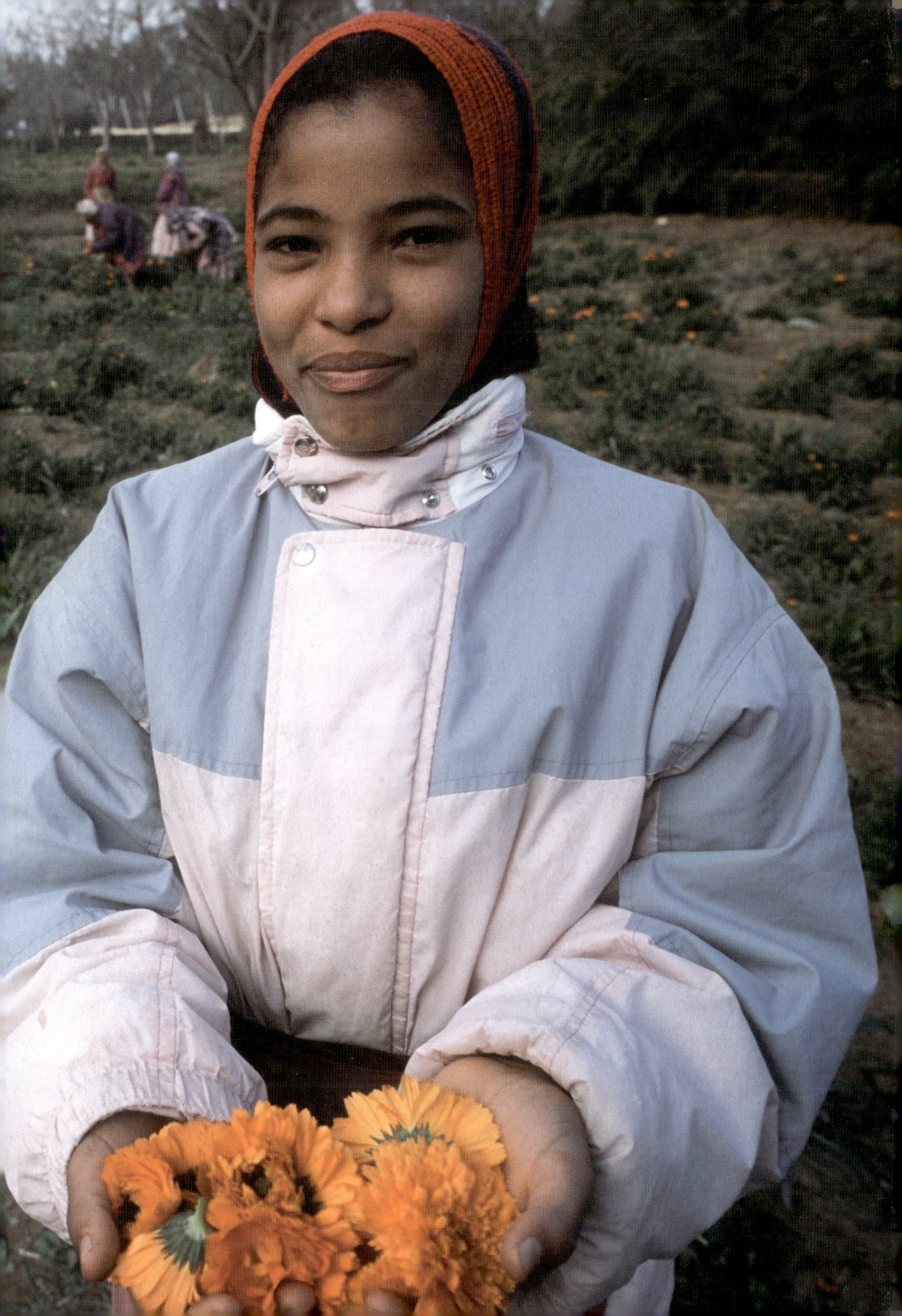

Contents

Foreword	11
The Vision	13

Part 1 The Story of My Life

1. Formative Years	17
2. Return to Egypt	55

Part 2 Founding a Desert Community

3. The Beginning	71
4. Prevailing	97

Part 3 Sekem — A Community Model for the Twenty-First Century

5. Economic Foundations	121
6. Education and Culture	163
7. Social Processes	203
A Chronology	225
Glossary	227
Photographic Acknowedgments	229
Index	231

Foreword

In 1977 the Sekem initiative was established with the setting up of agriculture, and over the years there followed commercial businesses, education centres and medical centres to complete the venture. Now Sekem has spread throughout Egypt. Many people, hearing about it, talk about the 'miracle in the desert,' or a 'dream too good to be true.'

In 2003, Sekem was awarded the Alternative Nobel Prize, the Right Livelihood Award as 'a business model for the twenty-first century,' combining social and cultural development with commercial success. The public recognized Sekem's activities as offering answers to significant questions of our time in the Arab world. How can an efficient economy, a healthy social fabric and a living culture develop together? How can partnership between west and east build peace and prosperity? What is the modern understanding and practice of Islam? Questions like this, and related issues, are commonly debated.

The aim of this book is to write about Sekem's complex relationships in more detail and to shed light on their background, to unite the story of the initiative with the story of the many courageous people who worked in it and to describe the 'miracle in the desert.' Fundamental laws of life have formed and accompanied Sekem, and that is why I would like to tell the story of this initiative using my own life story, which merged into the story of Sekem. At the same time I have a deep desire to thank everyone who was involved in creating this book.

First I would like to thank my wife Gudrun for her unending patience, care and love. She came to Egypt with me and actively supported Sekem from the start. She was able to take my daily worries off my hands, and through this created the space for implementing the Sekem idea. Her help with the manuscript has considerably contributed to its present form.

I would also like to thank my son Helmy. He has managed the economy of Sekem through its many ups and downs with incredible sureness of touch and strategic far-sightedness. He was able to inspire many people for the idea through his humour and positive attitude, which have come to influence markedly the content of this book.

My special thanks go to Barbara Scheffler, who awakened my memories with her patient persistence. She revised the entire body of information and created the structure for this account.

I thank Jens Heisterkamp for his spontaneous offer to edit this book. Working together with him helped me to bring order into my thoughts and ideas.

Thanks also go to Konstanze Abouleish and Hans Werner, who gave me valuable inspiration and insight and who lovingly accompanied the development of the book.

Throughout my entire life I have been fortunate to have people around me who have advanced, supported and loved me. This has filled my life with joy and helped me to develop. Unfortunately it is not possible to mention all the people to whom I owe thanks — but nonetheless my gratitude towards all those who have accompanied me and created the basis for this life story is infinitely great.

Ibrahim Abouleish
Sekem, August 2005

The Vision

I carry a vision deep within myself: in the midst of sand and desert I see myself standing at a well drawing water. Carefully I plant trees, herbs and flowers and wet their roots with the precious drops. The cool well water attracts human beings and animals to refresh and quicken themselves. Trees give shade, the land turns green, fragrant flowers bloom, insects, birds and butterflies show their devotion to God, the creator, as if they were citing the first Sura of the Koran. The humans, perceiving the hidden praise of God, care for and see all that is created as a reflection of paradise on earth.

For me this idea of an oasis in the middle of a hostile environment is like an image of the resurrection at dawn after a long journey through the nightly desert. I saw it in front of me like a model before the actual work in Egypt started. And yet in reality I desired even more: I wanted the whole world to develop.

I thought long and hard about what to call this project which I wanted to implement following this vision. Because of my interest in ancient Egypt I knew that at the time of the pharaohs there were two different words for the light and the warmth of the sun. And the sun also had a third element attributed to it: Sekem, the life-giving force of the sun, with which she enlivens and permeates the earth's entire being. I chose this name for the initiative I planned to start at the edge of the desert. This book tells the story of how it all developed.

Part 1

The Story of My Life

Opposite: Ibrahim, aged twelve.

1. Formative Years

Early memories

I always loved it when my grandfather came to visit us. I would be allowed to accompany him to the nearby mosque to pray. We stepped out of the house. Morning mist lay over the gardens and fields, through which the sun shone as white as my grandfather's robe. He took my hand and I felt his warmth and protection as we walked silently through the morning's stillness. He had plenty of time for me and there was no need to rush while we walked. We stopped at orange trees heavily laden with fruit, inhaled the scent of roses and watched happily while a butterfly sucked unselfconsciously at the nectar of a

Ibrahim's maternal grandfather and grandmother.

marguerite. My grandfather listened to all my childlike questions and found comprehensive answers for me which were deeply gratifying. He sat down beside the bright white flower with the dancing butterfly and took me on his knee. I leaned back against him, enjoying his gentleness. The butterfly opened its colourful wings and flew off the white blossom up into the blue sky. We both followed its flight for a long time.

After visiting the mosque I accompanied my grandfather to the bakery where we bought round flat loaves for the whole family. I breathed in the warm aromatic air, took the big round *fetta* into my arms and carried them home as if they were holy relics. My mother came towards us through the garden and I gave her the shining sugary breads. Then I ran off, climbed into the huge sycamore tree and lay swaying on its branches.

My grandmother also came to visit us with my grandfather and helped my mother do her sewing. As evening approached, she would put her sewing things away, fold her hands and sit back. This was the moment I was waiting for. I crept onto her lap and asked her to tell

1. Formative Years

me a story. I knew she could tell the best stories in the world in the most exciting fashion. I did not necessarily want new stories; I could have listened to the same story a hundred times without getting bored. My grandmother used to live in Menja, the city of Akhenaton. She liked to tell me about my grandfather, who came from Morocco and was a cloth trader in Cairo. My mother's childhood and later life was shaped by my grandfather's religiousness and the quiet friendliness of my grandmother.

My father's grandfather, who came from Galilee, was a wealthy cotton trader in Egypt. My father's mother originally came from Syria. I was my grandparents' first grandchild and they showered me with all their love. Although my sister Kausar was born two years later, and then every two years four further siblings followed, Mohammed, Hoda, Nahed and Mona, I always felt that my grandparents, and especially my grandmother, were there for me alone. My grandmother told me that my mother did not have enough milk for me after my birth and that she, as the older woman, had received milk and suckled me. Then I cuddled up to her and enjoyed her radiating warm life force.

I have two birthdays: the day I was actually born in Mashtul, and the day that is still written in my passport — April 27, which I thought was my real birthday up until my adolescence. My father carefully documented all daily events in a notebook like a diary. Once, while I was watching him I suddenly asked: 'Did you also write on the day that I was born?' He took out his notebook from 1937 and read out the entry about my birthday, written on March 23, the 10th of Muharram 1356 according to the Islamic calendar. After asking him why I had always thought my birthday was a different date about a month later, he explained it was usual to register births some weeks, sometimes months later, and then this date of registration was recorded as the actual date of birth in the passport.

My aunt Aziza had a wonderful garden with guava, mango, orange and pomegranate trees, bougainvillaea, roses and hibiscus. Whenever I was allowed to visit her for a few days she gave me fruits and fresh honeycomb to eat. She would stand beside me smiling, delighted by how much I was enjoying it. Early in the mornings we would get up and go to the dairy, where a goatskin hung from the ceiling by two ropes. She would pour cream into it and started shaking the skin, with me helping vigorously. We watched the milk curdle slowly until it had turned into lumps of butter. Then I would go back into her garden to play. Someone always kept an eye on me.

There was a well in this garden. Whenever I went anywhere near it, my attendants would get a scared look on their faces. One day during the midday heat I was all alone, I went to the well and bent down to gaze into its damp, dark depths. Suddenly I lost my footing and tumbled in; falling down thousands of metres — or at least that is how it seemed to me, as I appeared to fall such a long way. In reality it was more like three or four metres. There was no water at the bottom, but luckily it was not completely dry either. I landed with a thud and remained lying there. A woman came running up immediately and hauled me out of the well. But afterwards I changed. I now perceived the world through different eyes and realized that there were also dangers in it.

Cairo

It was the Sabet family, German Jews who moved to Egypt from near Stuttgart, who introduced artificial fertilizers to Egypt. My grandfather became a partner in one of the major artificial fertilizer trading companies of that time in Egypt. Initially my father worked with my grandfather, but later he established his own companies, which I will describe in more detail later. So we moved to the city when I was four years old. My parents showed me around a large, light-filled house with lots of rooms and many huge boxes filled with valuable dishes. Unfamiliar people brought beautiful furniture, varnished to a shine and smelling sweetly. My parents had decided to furnish the house in European fashion. My mother received visitors for tea in her magnificent drawing room. Berta Sabet, the wife of one of my father's business partners, often came to visit us. She wore posh hats and smoked using a long cigarette holder. My mother loved this elegant lifestyle.

We moved house repeatedly during this time and, and by the time I was six we were living in a flat on the third floor of a light-filled, beautifully decorated old house with many huge, and it seemed to me, high-ceilinged rooms. Many Jews lived in our neighbourhood. Our neighbour Rahel did not like seeing me and my sister skip down the steps every morning to our Christian school, St Anna's kindergarten. 'Ugh,' she would say, 'I don't understand how you can go to such a terrible school.' On Saturdays our neighbours would often come to us children and ask us to light their lamps for them, because they were not allowed to work on the Sabbath. In return we were given biscuits as a thank you.

Ibrahim's father, born in 1916, aged about 50.   *Ibrahim's mother, born in 1920, aged about 35.*

One day I was at home alone when the doorbell rang. When I opened it I suddenly stiffened. Then I ran to our balcony and screamed as loud as I could. I was only calmed when the house residents talked soothingly to me, and my mother came to hug me. 'Ibrahim,' she whispered, 'what's wrong with you? That was only old Mohammed.' Mohammed, the watchman of the house, lived in the basement with his wife. But I had never consciously seen him because he usually worked in the cellar. Mohammed was a terribly ugly, hunchbacked old man with shaggy hair, red eyes and big, yellow-brown teeth. He seemed like the devil incarnate to me as he stood at our door. Although my mother took me to see him afterwards, and I could look at him calmly while pressed tightly to her side, it took me a long time to overcome my reserve and only later was I able to hesitatingly give him my hand.

Ramadan — a wonderful month, the best time of the year when we children received small lanterns called *fanouz*. We opened a small door in this wonderfully shaped, colourful glass vessel and lit the candle inside: suddenly the room was transformed by the beautiful play of colours! My sister Kausar and I carried the lights through the streets and sang songs in front of other people's doors. I sang loudly and fervently as we got cake and other nice things for our performance.

Ramadan meant that we had lots of visitors. After sunset the tables were set abundantly with delicious sweets and in the evening a storyteller came with a *rababa*. He brought this instrument with him, which had two strings and was played with a bow. I sat at his feet and absorbed the images of the myths and epos, for example the story about the bold rider, who rode through the desert with his horse to save his friend. I sat in trepidation wondering whether he would succeed despite all the challenges facing him.

Ramadan — this was also the time of all the stories my mother told us about the Prophet. We listened reverently and in admiration to accounts of the Prophet's suffering and endurance, to how intelligently he answered questions, and to how much confidence he had in people's ability for freedom. The image of an admirable man was created in my soul: very gentle and wise, very strong and resolute. My mother and grandmother also told us about the Prophet's disciples. We did not want them to stop telling these tales, and so they sat with us till deep into the night. Even stories we had already heard were experienced again with excitement.

A great festival, Bairam, ends the month of Ramadan. We had a new outfit, new shoes and a new robe for this occasion. Everything smelled so good. My mother put all the wonderful things in our wardrobe and we were only allowed to wear them on the day of the festival. There were also other surprises. The women baked huge amounts of biscuits covered in icing sugar. We children sat around the kitchen table and enthusiastically helped our mother cut out small cakes which were filled with date puree. We placed them onto baking trays which were layered and numbered and then our maid carried them to the bakers on her head. Our kitchen was filled with a wonderful aroma when the baking returned. My mother sorted the biscuits into small boxes and I had to distribute them amongst the poor, which I liked doing. 'My mother greets you!' was all I said. The people were happy and expressed their gratitude.

In Cairo there are Turkish baths decorated with beautifully coloured glazed tiles. As a five-year-old I was fascinated by what lay hidden behind these magical walls. My parents never went there. After much pleading a neighbour took me with her one day. We got undressed and entered a room dripping with water and full of steam with many naked women sitting in it. I was so shocked and terribly self-conscious that I ran home as fast as I could.

Once when I was alone in our big house I leant out of the window on the first floor to watch some boys playing football on the street below. They skilfully kicked the ball up to the marked goal and then took a shot at it with tremendous force. Full of admiration I watched their skill. Then one of the shots missed its mark, and the ball landed on our roof instead of in the goal. Like all houses in Cairo, our roof was flat. The boys ran up to my window and pleaded: 'Ibrahim, please get us our ball off the roof.' I climbed up to the roof and found the ball, but did not want to throw it down — that did not seem dignified enough – so I decided to kick the ball. I carefully placed it on the edge of the roof, ran up to it and went flying off the roof with the ball. I hit the ground hard. My whole body hurt. The boys ran off in shock to get big Ali, the tailor from the neighbouring house, who carried me to hospital in his strong arms. I was in quite a state and had a broken foot. I swore never to play football again!

My father enrolled me in a French school, where I learnt French for three years. Later my sister Kausar also went to this school. Soon we both realized that we could speak a 'secret language' which nobody else in our home could understand. My mother suffered greatly from this and succeeded in having me transferred to an Egyptian school. Kausar was allowed to stay on until her school-leaving exam. Each of us children had their own room in our house. Kausar's room was always neat and tidy, and I also made sure that everything around me was kept in order. Because of this we started locking our doors to stop our younger brother Mohammed from entering.

My father established his own business and started becoming interested in industry when I was about nine years old. He set up a soap factory and a confectionery factory in which he produced the famous *halva* made from honey and almonds. The place he chose for the factory was characteristic for my father. Shortly before establishing his business there was a terrible attack in the area heavily populated by Jews. An ice-cream van parked in one of the streets was blown up by a bomb planted by extremists just as the van was surrounded by children. My father built his factory on the exact spot where the bomb had left its devastation to show that such things should never be tolerated.

My father set up two large companies employing many people.

Every day after school I changed my clothes and went to the factory. I was greeted by the smell of soapsuds boiling in huge vats. The workers waved me over, handed me one of the long poles and let me stir the thick suds from above. Then the hot lye was poured into frames lined with greaseproof paper to dry overnight. At this point the scent was added following a secret recipe: real rose oil, oil of lemon and orange oil — it smelt wonderful! The next day the 2 × 2 metre blocks were cut into 60 × 60 centimetre pieces with kitchen knives and string, and then they were lifted onto a table. There I stood and pushed the large squares of soap through stainless steel wires attached to the table at right angles. First the soap was sliced into long thin slabs, and then these slabs were again cut into smaller bars of soap. One had to be extremely careful. It was easy to cut a finger, and then the soapsuds stung in the wound. The finished bars of soap were again left to dry and stamped with the company name. The soap was called 'Al Doctor,' written in a half-moon shape. I liked letting the soap glide through my fingers, enjoying its silky feeling.

There was a carpentry business beside the soap factory where they made wooden boxes for the bars of soap. I enjoyed working there and learnt a lot about working with wood. A most enjoyable moment was had in the packaging. Colourful tissue-paper, ribbons and the beautiful writing of the commercial name 'Al Doctor' perfected the product. I would have loved to keep all the splendid packages to give away. The same was true of the products of my father's other factory, where he made confectionery. From the packaging process I learnt the importance of matching the colourful shiny paper to the smell and taste of the sweets.

There was also a haulage business attached to the factory. Handcarts and donkey carts took the goods to the market, the storeroom or the shops. The area around the factory was a Jewish quarter, and the synagogue was right next to the factory. We were the only non-Jews there. Because of this I grew up to be tolerant. A Jew meant a friend. When I left the factory in the evening a loud cry would greet me from a corner: 'Ibrahim, please come and take me for a ride!' Dudu was a handicapped Jewish boy who was actually called David. He waited patiently until I appeared. Then I sat him on one of the handcarts and pulled him through the alleyways. Dudu shouted with joy. It was so easy to make him happy with a joke or a bit of teasing! He never wanted to let me go. Other people watched us playing, and soon parents started bringing their handicapped children for me to pull in the cart and play with them. Once they got too heavy I harnessed a donkey to the cart to pull it through the lanes.

Our tailor, who could cut out and sew the most perfectly fashioned garments by hand, was also a Jew. I was always proud to parade through the streets wearing new clothes made by him! Of course I also wanted to show my father my new suit. On the way there a boy called out: 'Ibrahim!' I looked around. The boy was coming towards me, but I did not know him. How did he know my name? With innocent eyes he told me he had been sent by my father to tell me he was waiting for me in the bank. So I turned in a different direction, with the boy accompanying me. He praised my new suit but suddenly said sadly: 'Did you know your new jacket has a stain on its shoulder? I'll try and clean it.' He starting rubbing and wiping it. 'I can't do it. Take it off, I'll quickly run inside and wash it with some water. I'll be right back.' I gave him my jacket — but he never returned and I stood on the road for a long time before realizing what had happened.

I had an excellent Arabic teacher at school whom I greatly admired. He was the perfect kind of teacher, correct and pedantic. His neat clothes, his writing on the blackboard, his eloquent speech filled me with enthusiasm and left a lasting impression in my soul of an honourable man. Abu Affifi was my idol, and when called to him I hardly dared enter his room. In the upper school we had a French teacher who became my good friend. He came from Morocco; spoke perfect French and only broken Egyptian. I refreshed my French from my first two school years and talked to him in his own language, which gladdened him. My gymnastics teacher was the regional champion and over the years gave us excellent training in apparatus gymnastics and floor exercises. I practised enthusiastically for competitions. I walked around our flat on my hands, and could also go up and down stairs, much to the merriment of my family.

My entire name is Ibrahim Ahmed Abouleish. Because Arabic does not differentiate between I and A my initials are 'AAA.' In school we were seated and called upon alphabetically. So I was always the first and always asked first. I would rather have been somewhere near the middle so that I could listen to others before answering difficult questions. But as it was, these triple A's were not always an easy fate to have.

Unfortunately I did not have a good Arabic teacher in high school, and lessons and tests were sometimes agonizing. Arabic grammar is very difficult. I was often marked: 'content good; writing could be better.' I was also not able to and did not like making speeches. The children from the countryside were much better at this because they were less self-conscious. So I concentrated on the sciences instead. I was so fascinated by what I experienced during laboratory experiments in physics and chemistry that I could never get close enough and ended up nearly sitting on the experimenting teacher's desk. I did not just want to learn by heart, I wanted to understand what I saw and experienced. I found mathematics quite easy because the teacher was good at explaining the concepts.

I had a beautiful, tidy study beside my bedroom where I could learn in peace. When I finished studying in the evening my mother would be sitting sewing in the living room. When I entered the room she would lower her handwork and look at me with her delicate smile: 'Would you like to tell me a bit about what you have been studying?' She could not speak English or French, nor did she know anything about physics, chemistry or higher maths. So I sat down and explained to her what I had just learnt, which also deepened my own understanding of the subject. She listened intently, asked questions and wanted to know every detail. Although she had so many children she saved this time of day especially for me. I developed the ability to teach others from an early age through this atmosphere of close affection uniting us.

Summer in the countryside

I always spent my summer holidays in the countryside in my home village Mashtul in the Nile delta about fifty kilometres north of Cairo. Two of my aunts lived there in houses far apart from each other. It took me hours getting from one to the other because I met so many friends on the way. 'Ibrahim is here again!' everyone said. 'Tell us about the city! Have you written anything new? Read it to us!' The village youths gathered around me and we laughed and joked together. I told them the newest stories or recited the poems I had written. Once they told me there was not enough water in the canal and the water had to be lifted into the fields with an Archimedes screw. I had never seen this machine or how it worked, as I had been shielded from farming work as a young child. In those days we had labourers who did all our manual tasks. Now I thought it was a miracle how the water was transported up from deep below, poured forth into the ditches, slowly trickled along and was diverted through an embankment onto the fields. 'Let me do it too!' I could have turned the handle for hours, amazed at the wonder of the machine. They also took me to thresh the corn with a *norak,* a board with iron discs that was placed over the corn. This was slowly pulled over the straw by a buffalo, cutting up the straw so that the grain fell out. I observed the proceedings and talked to the workers to find out about their way of life. After some time they also started talking about their worries and needs. I wrote everything down in a small booklet, including the name of the person and a short note describing their needs. I was interested in people, and I am sure they felt my affection. I found it easy to make contact, especially with the poor and sick. I kept the booklet on my desk in Cairo, and thought about what I could bring back to them on my next visit as a surprise: soap, cloth, clothes, shoes and confectionery. I wrapped up everything beautifully and hid the packages in my suitcase away from the prying eyes of my siblings and parents, as it was my secret. I did not want to give the things to the people myself. So late at night I wandered through the village from house to house and threw my parcels through the open windows. Then I hid to see what happened. This gave me quiet pleasure.

Nobody knew about my actions apart from my mother, and she helped me get the presents together. Only once when I asked for one of her beautiful dresses she said gently: 'Ibrahim, isn't that a bit much?' My grandmother found out about my deeds when she surprised me whilst I was unpacking my suitcase. She simply asked: 'Have you got anything for me?' So I told her everything. I was very grateful that neither of them passed on my secret, for they knew me well. I was an extremely happy and humorous child, but also very proud, dangerously proud, as my mother said. There were certain things I could not be told as they hurt my pride. When I was hurt, I withdrew and stopped talking, which is still the case today.

One of the people living in the village needed to go to Cairo to the doctor, because he could not be treated in the village. I found out how much it would cost so I could organize the money for the doctor's bill and hospital stay. In those days I hardly had any relationship to money; sometimes I would use my own savings to help the people, which then became known. One of the five pillars of Islam is giving alms to the poor, known as *zakat.* Because of this my relatives gave me money to distribute to the needy. I looked after this money carefully and wrote down who gave me how much money, and whom I passed it on to. Because of this I was called Ibrahim Effendi or Ibrahim Bey from an early age, which showed the respect that people felt towards me.

Adolescence in Egypt

The years between 1952 and 1956 were of great importance for the political future of Egypt and were accompanied by unrest in the population and demonstrations in the streets of Cairo. In 1952 officers of the opposition overthrew the Egyptian King Faruk. One year later the Republic was founded and in 1954 Colonel Abdel Nasser became President. But the young Republic remained strongly under the influence of the British, who had their troops concentrated on either side of the Suez Canal. In the same year, Nasser was able to secure a contractual guarantee from Great Britain to remove its troops from the Suez Canal and thus out of the whole of Egypt, which only actually happened in 1956. Up until this time the Egyptians protested against the increasing corruption and injustices of the British rulers. I closely experienced this time of political unrest in Cairo and heard about the clash between the Egyptian police and the British army, where 100 police were shot at the Suez Canal. The people took to the streets to protest. In 1952 Cairo was in flames as many foreign and especially British shops were set on fire in protest. High school children were given days off school to take part in the demonstrations.

Paternal grandfather with Uncle Essad, around 1940.

1. Formative Years

I had a few close friends who were also interested in social and cultural topics. We would go rowing on the Nile when we did not have to go to school. I also did many bicycle trips with these friends throughout the whole of Egypt, during which we got to know our country very well. Every weekend we had off would go to El Faiyum, Port Said or Alexandria. Once I organized a bicycle trip to El Faiyum for five adolescents. It must have been during the holidays. I informed myself about the trip and organized everything before leaving. We wanted to travel there on day one, sightsee the next day, sleep in the youth hostel and then return. It was a remarkable day as the oasis El Faiyum is wondrous. We saw the huge waterwheels at the Yussuf canal, which leads from the Nile to Lake Karun. We rode for miles beneath tall date palm trees along cotton and sugar cane fields. We also visited the death temple of Amenemhet III, built like a labyrinth. I was able to pursue my cultural interest in ancient Egypt with these friends, as we were all similar: we did not drink or smoke nor sit in coffee houses.

On our way back we were hit by a terrible sandstorm. We covered ourselves with our blankets, holding them onto the handlebars while the storm drove us along the road with incredible speed. Suddenly we heard a cry. One of our friends had fallen and his bike was broken. It was impossible to fix the bike during the storm. The sand whipped against our naked arms and legs like pinpricks and our eyes burned. As the others had already rode on, I let him sit in front of me on the handlebar, held on to his bike with one hand and steered my bike with the other. We sped on for hours, propelled by the storm, back to Cairo. My friend still tells me today how he was shaking throughout the entire hazardous ride.

I had many such experiences at that time which tested my limits. Rowing on the Nile tried my willpower and endurance. I also climbed the snow-covered mountains of the Sinai. These were challenges I chose myself. Although I wanted to get to know these mountains, I also viewed these excursions as a chance to test my limits.

Certain individuals in my family influenced me greatly. One of them was my father with his business, then my mother and grandmother, but also three uncles: uncle Kamel, my mother's brother, had a transport company and took me to faraway places and villages. There I got to know people who spoke and dressed differently and ate different foods than those in Cairo. Uncle Mohammed, my father's brother, was an extraordinary cheerful person. His wealth was all inherited, but he had never done anything with it. He married seven or eight times, and liked to spend time in the theatre or opera. Because he was a connoisseur we knew that whenever Uncle Mohammed came to visit he would bring something interesting and beautiful. In those days, all the time, even to school, we all wore a fez, called a tarboosh in Egypt, a red, tall Turkish hat with a twisted tassel. When my uncle came to pick me up in his car he would hoot the horn I would call out: 'I'm coming, I've just got to get my fez.' He would always shout back from below: 'Come on, have my fez and forget your tarboosh!'

In those days there were streets in Cairo lined with art and handiwork stalls. We strolled up and down these lanes savouring the sights, and he would ask: 'Do you like that? Look at how beautiful that lady is. She moves so gracefully!' Sometimes he also criticized things he did not like, and I learnt to see a whole new side to life through his eyes and judgment. Later I realized he did not necessarily have much taste, but he was someone who awakened my interest in music and theatre.

A further uncle, also called Mohammed, was a university professor and had a library in his house. I felt great reverence towards him and did not dare speak to him as freely as to the others. I entered his house with veneration. He asked me highly philosophical questions, and I had to think through the answers carefully so as not to embarrass myself. Friends visited this uncle to discuss philosophical, religious and historical themes. They conversed in the refined, classic Arabic language. I sat slightly apart, following their erudition with astonishment and admiring their ability to think such thoughts. I always hoped Uncle Mohammed would ask me to bring tea or water. None of my friends could understand what drew me there or how I could stand the endless discussions. The other members of my family tried to avoid this circle of scholars and even complained about the 'gibberish' which so fascinated me.

Sometimes Mohammed would give me a book out of his library, or I was allowed to choose one. One day I came across Goethe's *Sorrows of Young Werther* in Arabic, which I wanted to borrow. My uncle was not very happy with my choice, but I was deeply moved and anguished by this book. Partly this was due to the fact I was terribly in love with Awatef, the daughter of one of our neighbours, at that time. Awatef had a beautiful voice and wrote poetry. When she read her verses out, I absorbed the pitch and every nuance of her voice, every movement of her eyes and hands. After meeting her I lay in my room with beating heart, and tried to express my love through composing my own verses. Awatef was two years older than I. Because it was not seemly to fall in love with an older woman in Egypt I kept my love a secret. But I think my mother knew how much I worshipped this girl. She sometimes invited the neighbour and her daughter Awatef to our house. We sat drinking tea and she talked about things. It was wonderful! If I had not see her for a week I would attempt to get my mother to invite her, or her mother to invite us. Awatef started her studies two years before me and studied natural sciences. I was quite envious that she was doing something I was not allowed to do.

Goethe is pronounced 'Gota' in Arabic, because there are no equivalent vowels in the Arabic language. My father once had a visitor from Germany. He was a tall, older gentleman. I told him: 'I know a German writer, he's called Gota.' 'Really? I have never heard that name before. What else do you know about him?' So I told him what I had read of his. He called out surprised: 'That's Goethe! You mean Goethe!' He taught me how to pronounce the umlaut. This educated man also told me about Schiller and the friendship between both writers. I avidly absorbed his words and after meeting him wanted to discover more about the German people and their writers. Everything I learnt about this country and Europe strengthened my longing for European culture. Art and science, the economic life, the rights of the citizens and their possibilities — I deeply admired it all. I had also seen parts of European landscapes in films and they appeared most beautiful.

The closer I came to my finishing exams, the more seriously I thought about going to Germany to study. I asked my father whether he would support my wish, but he disapproved vehemently. Every time I brought the subject up he was displeased. One time I heard him talking to my uncle about this young man who persistently wanted to study in a foreign country; I did not realize he needed a successor for the factory and that one did not need a university degree for that. I was dependable and hardworking and should not leave him, my father thought. So I could not hope for my father's support.

My mother had tears in her eyes every time I mentioned this subject. 'The country you want to go to is so far away, Ibrahim; I would not be able to see you for so long — and what will happen to father and his business? Don't do this to me. Stop thinking about it!' I still kept talking about the subject; I did not want to sadden her, but hoped she would be able to understand me when I left — because deep in my heart I had already decided. A few times she said she was happy that I was doing what I really wanted to. This strengthened me. Then again she would plead with me not to go.

It was also difficult for me to leave Egypt. I would not see my beloved Awatef anymore. How could I forget my love? This was the first time I had the chance to practise not getting too attached to things. Maybe I had some premonition that this was only the start!

I had saved some money which was enough for a single ticket, and secretly hoped that once I was in Germany my parents would send me more. My behaviour was very risky considering how family ties in Arabian countries work: the father is the undisputed head of the family and decides everyone's fate. My mother had a double burden to carry: as well as her personal pain at our parting — we were connected by a deep love — I forced her to be a mediator between me and my father and hoped with youthful unconcern that she would be able to pacify him and persuade him to give me the necessary financial support. I owe her deep thanks in retrospect, as she was able to bring him round by her gentleness, paving the way for the future I had decided on.

Departure from Egypt

I had written to universities in Europe and they told me I could come. In Egypt, every young person planning to study abroad needs to report this to the government and apply for a passport. My father was supposed to leave money at an administration office, from whence it would be transferred to me abroad. Because I was still underage, I needed my father's signature for all the immigration formalities. I could not leave without them, but how could I procure his signature? I pondered for weeks. Then one of the civil servants, who had noticed how I was suffering, said to me casually: 'It's easy!' — 'What?' — 'The Foreign Office does not know what your father's signature looks like!' — 'Yes, but what if they ask him?' — It seemed like a criminal act to me. For three months I struggled with myself and fell into deep despair, until I found the courage to fake his signature. Theoretically my father would have had to sign in front of them.

During my trips to the authorities I met someone whose father had sent him to Graz in Austria to study a few months before. He told me the country was very beautiful and gave me his address. So I knew my destination, even if I had to look it up on the map first, as I did not know where that city was located. My friend Shauky, who would have liked to study in Germany, accompanied me to Alexandria. It was winter and raining. Early one morning, my mother was still sleeping; I opened the door quietly, gave her a kiss on her forehead and said: 'I'm going now!' and stole out of the house. Later she wrote to me saying she thought I was only going into town for the day. If she had known I was leaving she would have given me a kiss and a hug.

I wrote the following farewell words to my father, in the hope that he would at least understand my decision. I was in a state of inner turmoil while writing this letter, eighteen years old and having decided to go against the strong family ties — much stronger in the Arabian world than the western world — and go my own way. Perhaps the clarity of the vision of my future expressed in these words can be attributed to this letter, which my father showed me 25 years later.

My dear father!

Peace and greetings be with you.

When I get back, if God wills, I will go to Mashtul, the village I have always loved and where I spent the best time of my childhood. I will build factories where the people can work, different work than they are used to from farming.

I will build workshops for women and girls, where they can make clothes and carpets and household goods and everything else that the people need. I know that transportation and communication means are very important, so I will let the road be tarred from the station to the village, and plant trees to right and left of it. I will establish shops that sell everything the people need, even a casino, like a huge market, but very tidy and clean.

I will build a large theatre on your grounds, where renowned artists can give performances for the people of my village.

Near the main road, which will lead to Aesbet el Barkauwi and Minia el Kamah, I will build a hospital which I will fill with specialists. In the village I will make a small quarter for the doctors and their assistants and teachers to live. I will need teachers as I want to build schools for the children, from kindergarten to high school.

Mashtul has got men and young people of higher learning, for example Dr Schuman (doctor), Ustaz Orabi (lawyer), Ustaz Afifi (teacher), Ustaz el Gohari (sheik), Ustaz Umara (engineer) and many more whom I am sure will enthusiastically help me establish my idea, so that the village of Mashtul can become a shining centre in Egypt.

Peace be with you.

In Alexandria I bought a ticket to Naples on a Turkish ship for 30 pounds. From there I wanted to take the train to Rome and then on to Graz.

As the ship slowly sailed away from the shore, I thought to myself: 'What have you done!' and my heart rent asunder. My friend waved from the quay, the huge ship slowly sailed out into the open sea, and I felt the connection to my mother and father, to all my relatives and friends and to the country of my birth painfully separating as the ship moved away. I was giving up everything I knew for an uncertain future in a country I knew nothing about and whose language I could not even speak. I also did not have much money with me, just enough for the journey.

I suffered achingly for days until I got to know some of the others on the ship and slowly calmed down.

Once in Europe, I admired the buildings and works of art in Rome. Then I travelled on to Graz via Florence and the Alps. North of Florence it began to snow. For the first time I realized what real cold meant and how unprepared I was for my undertaking. I bought a sweater which some Italians were selling in the train to Graz.

Studying in Europe

Once I got to Graz I went to see the friend who had given me his address in Cairo, and he welcomed me warmly. He soon found me a room in a beautiful area outside the city. My landlady had been a nurse in England and could speak English. I wrote to my parents immediately, not forgetting to mention they could deposit the money at the administration office. My mother wrote back to me how she had suffered since I left and that she found the separation difficult and was always in tears. And whenever I read her letter I also had to cry. Despite her pain at my departure she was the one to take the money to the authorities and organize everything for me.

During the following days I discovered how many different departments there were in the university. Of course I wanted to study natural sciences. If it had been up to me I would have studied medicine, but I could not do that to my father. In his eyes, I should do something 'worthwhile,' something he could use in his business. And I wanted to help him and at least meet him halfway. So I decided to put my own wishes to the back of my mind for the time being.

1. Formative Years

Once in the university, I got in contact with the club for foreign students. When they discovered I could not speak a word of German — I only owned I handbook called *Teach yourself German* — they persuaded me to learn German first. The students in this club were mainly children of rich parents with lots of money, who spent their time playing cards, smoking and going out with girls. I set about systematically learning this foreign language following my own method. Every day I learnt a certain amount of vocabulary. My fellow students complained about the difficulties of learning the German language. Even after several years in the country they could not fully understand the lecturers. One of the other students, who had still not written a single exam after five semesters, decided to dare me. He said I would not manage to win, and we made a bet in front of many witnesses that I would finish studying before he did. Because of this I discovered what he was studying: technical chemistry, which I had never heard of before.

The next day I signed up for this subject and absorbed everything with concentration. But it was not like studying, it was pure horror! I was in the laboratories from eight in the morning until eight in the evening, and was also studying physics, mathematics, crystallography, geology and more on top of my main subject. My fellow students took eight to nine years to complete this degree. After only three semesters I was able to teach the foreign students, helping them to translate and preparing them for exams I had already sat. In those days there were no official exam dates, each student went to their professor and asked to take the exam once they felt ready. But who would voluntarily choose to sit an exam! Everyone put it off.

At the start of my studies I went home for a short visit at my mother's request, on a ship from Genoa to Alexandria. In Egypt I was welcomed with open arms. While I was packing my bags shortly before my departure, my mother brought me a heavy suitcase, full of wonderful sweets she had baked for me. We argued about how much she had given me, but then she said I could always share them out, so I took them along.

Adel's father, the boy who had given me his address in Graz, asked me for a small favour which I granted him. Just before my departure from Alexandria he came to the pier flanked by two soldiers and gave me two huge, heavy jerrycans. 'What's in them?' I asked, horrified. — 'Old cheese and black honey, sugarcane molasses,' he said, adding that his son in faraway Austria was surely missing it as he used to love it so much. I dragged the luggage to my cabin, but the cheese developed such a strong smell after only a few hours that I had to lug them back on to deck. There they stood, those dreadful jerrycans, and I debated with myself whether I should carry all this stuff from the ship to the train and then from the train back to my room. — No! — I gave the cans a good kick and they sank into the Mediterranean. My friend had a good laugh when I told him, and I gave him some of my mother's sweets as compensation.

The ninety-nine Names of Allah

During my first years of studying I often felt quite lonely in the unfamiliar city. I had given up my home country, and life in the club did not suit my way of life. So I put all my energy into my studies. I had a few friends who were interested in music and we went to the opera or a concert once or twice a week in the beautiful Stephanie hall in Graz. But one thing remained from my childhood and adolescence which filled me with strength, like a strong current from my past, carrying me through lonely times: my religion. The Koran accompanied me through my daily meditations and praying times, which I had kept throughout childhood.

In the Sura 'The Cow,' Allah is praised with the words: 'Allah, there is no God but him, the living, the eternal.' I trusted in this constancy now. 'He does not have sorrow or sleep. Everything on earth and in heaven belongs to Him ... He knows what lies before them and behind them, while they know nothing of his being, apart from what He lets them. His throne encompasses heaven and earth, and it is not difficult for Him to keep them. He is the highest, the greatest. ... Allah, there is no God but Him. His names are the most beautiful' (Sura 20, 255).

Islam is a monotheistic religion and accepts Allah as the one and only God. But he receives different names, ninety-nine in all, which the Muslim can meditate upon. These ninety-nine names hung in my study opposite my desk, printed on a beautiful piece of leather, and they are still part of my life today. In one verse the Prophet says that we should strive towards attaining Allah's qualities: 'The ideals of Allah are the highest in heaven and on earth.' I meditated on these ideals and they gave me the strongest support in those years of studying in Graz. I contemplated the names in my own way. I thought: 'He is called the patient one — I will practise patience. He is the knowing one — I will become knowing. He is the experienced one. He is the one and only. He is the strong one. He is the merciful one. He is the forgiving one.' And every time I was meditating on one of these ideals I would find myself in a situation where I could practise them; for example, to forgive instead of flying into a rage, which would have been more suited to my character. And I was often confronted with experiences which brought Allah's qualities to mind: 'Allah is the patient one,' so I practised patience. Because of this, these years became years of inner exercise, although throughout my life I have always experienced and thought of myself as a practising person. I never saw any of the difficulties I encountered as an attack on myself. It was always a chance to practise self-development.

This shows how my way of experiencing religion differs from that of my fellow humans. Nowadays I live consciously and try to impart what came naturally to me then: Allah is not a God who sits alone and unattainable on his throne in heaven without contact to his beings. Through my inner exercises I tried to establish a relationship to him. Because of this I do not want to be known as a religious person, but rather as a striving, practising person. I had a goal, an ideal — Allah's qualities, his ninety-nine names. When a situation became unbelievably difficult for me I could still see how small I was in comparison to his names. In this ways my difficulties became bearable. During my stay in Europe I got to know the Christian religion and its teachings of the Trinity. This is unacceptable to the Muslims, as they perceive Christians seeing the one and only God personified into three Gods, and Allah forbids further Gods beside him. The Sura 4, 116 says: 'Allah will not forgive those who serve other Gods besides him. He will forgive all other sins, if he chooses to. But whoever serves another God besides him has erred greatly.' In reality, Christians do not think of further personifications of the God in the Trinity, but rather see three different qualities revealed in the Trinity. I soon noticed how much misunderstanding this led to between the two religions. In later years I discovered the qualities revealed by the idea of the Trinity in Allah's ninety-nine names. The God Allah encompasses three qualitatively different areas with thirty-three names each. Muslims are requested to consider these names as an example. 'To BE is the highest ideal.' As these ideals had and have played such an important part in my life, I would like to notate the three separate qualities:

The ninety-nine Names of Allah

The One	The Light	The Judge
Whose Being is One	The All Watching	He Who Contains Everything
The Holy	The Vast	He Who Contemplates Every Detail
The First	The All Knowing	The Witness
The Last	The King	The Most High
The Visible	The Wise Adviser	The Most Great
The Recognizer of the Supersensory	The Glorious	The Experienced
The All Hearing	The Preserver	The Great One
The Wise	The Honourable	The Rich One
The Most Glorious	The Eternal	The Determiner of Fate
The One Who Elevates Himself	The Representative	The Lord of Creation
The Creator	The Giver of Life	The Fair Bestower
The Sovereign Lord	The Death Enabler	The Generator
The Benedictor	The Expander	The Maker
The Originator	The Constrictor	The One Who Registers
The Repeating One	The Expeditor	The Almighty
The Fashioner	The Delayer	The Protector
The Evolver	The Beneficent	The Firm One
The Hearer	The Destroyer	The Avenger
The One Who Prevents Disaster	The Bestower of Honour	The Enricher
The Uplifter	The Evoker of Humility	The Strong
The Abaser	The Expert	The Prevailer
The Merciful	The Truth	The Justness
The Source of Peace	The Alive	The Good-Natured
The Loving One	The Forgiving One	The Loving
The Initiator	The Patient	The One Who Feels Responsibility
The Infinite Inheritor	The Bestower	The Pardoner
The Beautiful	The Composed	The Generous One
The Appreciater of the Heart	The Exaltor	The Leader Towards the Right
The All Seeing	The Sympathetic	The Superior
The Thankful One	The Compassionate	The Guide Who Prevents Repeating a Mistake
The Affectionate One	The Resurrector	The Forgiver
The Consistent One	The Trusted Friend	The Honourable and Generous

Starting a family

When walking home from a night out at the opera I was often accompanied by an older man with a tailcoat carrying a horn. Kajetan Erdinger was a horn professor at the music school in Graz. We got talking, and one day he invited me home to meet his wife and three daughters, who all welcomed me warmly. Mrs Erdinger, who spoke perfect high German despite coming from lower Austria, noticed my broken German and offered to give me some lessons. Kajetan Erdinger was a man with a strong connection to nature. He had an aviary in his garden with the most colourful birds from all over the world. He was able to copy their different calls with his lips or by using stones while we waited eagerly for an answering call. He also took me to see the kennels where he bred shepherd dogs. Then he showed me his carrier pigeons and taught me how to treat them. At the weekends we went for long walks in the woods around Graz. We both had a rucksack on our back and walked silently beside each other through the solitude. Now and again we would stop. He gathered herbs and medicinal plants for his wife which she dried to make teas. Kajetan Erdinger knew all the names of the plants and their healing properties; I could only marvel at his knowledge. He opened my eyes to nature, which is so different in Egypt, and gave me a sense of its consistency and its mysteries.

The oldest child was Gretel, and then came the twins Gudrun and Erika. Gudrun, the elder of the two, was a beautiful, lively girl with a strong will and grew up like a son to her father. He called her Gundel. She helped him a lot with his chores around the house. She was sixteen years old at that time, the most active and social of the girls. I fell in love with her. I often watched her walking to the bus stop with her twin sister from the window of my student room. Her gait, in comparison to her sister's gliding walk, showed her strong, firm personality. Both sisters were doing the teacher training course. Once we got to know each other better she would often visit me in the laboratories and sometimes bring me a sandwich, which was always a lovely meal. Sometimes she came and said: 'I am not going to my piano lesson now. You can dictate to me what you want to write' — which naturally I loved doing.

Once, in summer, I was allowed to accompany the mother and her three daughters on a journey to lower Austria. Those were wonderful weeks spent among beautiful wine hills. We walked over green, blooming meadows, climbed mountains and viewed all the beautiful churches in Krems. We laughed a lot, but also sang. Mrs Erdinger sang in a choir and had a wonderful voice. The whole Erdinger house was filled with music. Whenever I visited I heard singing or the sound of a musical instrument coming from the rooms. Gudrun played the piano and her sister the violin. All of them had lovely voices. In the evening or Sunday afternoon Mrs Erdinger lit candles and the whole family gathered around the grand piano in the sitting room to make music together. I thoroughly enjoyed the joyfulness and vibrancy in the house.

Kajetan Erdinger often took Gudrun on his walks through the woods. So we were able to get to know each other more thoroughly. I saw her as a trustworthy, strong woman and had one wish: to keep her by my side forever. At the end of the year I asked her parents for her hand in marriage. Both of them raised their hands above their heads and cried: 'But you're still a student! How will you support a family?' — 'Soon I'll be finished and then have a career. You know I'm a hard-working man!' And I mentioned a hundred further reasons to persuade them we should have the wedding soon. Eventually they relented under the condition that we would have a church wedding, as they were Catholics. I did not mind as long as the wedding was soon. Before the wedding I had several long and interesting discussions with the priest. This extraordinary happy and uncomplicated man's beliefs were not at all fanatical. He accepted my wish to remain a Muslim, and used our discussions as the chance to find out about Islam. I also received important information about the Catholic faith. I remember we laughed a lot during these discussions. For example, when I said that we would naturally bring up our children to be Muslims, he asked: 'And who do you think will make the children Muslims? You as the father, or your wife, who hardly knows anything about Islam? She will bring them up to be Catholics!'

We got married in November in Graz with a big celebration, and then went to Wachau in lower Austria for our honeymoon. Unfortunately my parents and brothers and sisters could not attend the wedding. But they had been to visit a few times from Egypt. Later, once our children were born, the two very different families were also able to meet in a friendly and heartfelt way. My parents were happy with my decision. My father, with extraordinary tolerance, took straight to Gudrun, and later to his two grandchildren.

We moved into a lovely new flat. Gudrun finished her teacher train-

Gudrun with Ibrahim around 1987.

ing course and was soon expecting our first child. Our son Helmy was born in the Eggenberg sanatorium in Graz. I received the news just after I had sat my state exam, which took place on the same day. Full of exhilaration I rushed straight to the hospital from the university in my black suit without getting changed and took my son into my arms. I remember how the nurses and doctors made fun of me as they thought I had got dressed up to greet my son. Two years later our daughter Mona was born, and I also came to see her in a black suit, because again I came straight from an exam. After that I was known as the 'noble father' by the sanatorium.

Gudrun lovingly brought up the children and cared for them day and night. She sometimes complained about me: that I did not have enough time for her because I was caught up with my studies until late at night. When Helmy was four and Mona two we drove to Athens by car and from there took a ferry to Alexandria to visit my family in Egypt for the first time as a family. We received a warm welcome from everybody.

My time as a student

Technical chemistry consists of many different subjects, each one of which could have been an entire subject on its own. They all interested me and I was inspired by everything. I always had the feeling the professors knew so much about something I knew only a bit — I wished I knew as much as them! When I asked an assistant a question, the answer was so full of knowledge that I was overawed and thought: If only I knew as much as the assistant! I had a persistent urge to educate myself, to learn any missing knowledge through books and to work on the subjects with the help of my lecture notes. I followed the lectures with so much concentration that I was able to remember the professors' every word. Afterwards I repeated the lecture from beginning to end, going over every detail: What exactly did he mean? What were the thought processes? I voluntarily rigorously schooled my thoughts. The effort involved was heightened by the fact I was studying everything in a language I had only just learnt: German, the language of famous poets and philosophers. I filled entire books with my notes so as to better understand the idea behind the subject. Then, when it felt it dawning on me, I would be excited and contented. I did not understand problems straightaway, but had to struggle with them, and often carried a question around with me for days on end. Because of this I would describe myself as a slow person, someone who has to digest things first and only slowly comes to grips with a problem.

I noticed my fellow students, whom I taught, had a completely different way of understanding. They found it more difficult to really understand the processes involved in the subject, and so they ended up learning most of it by heart. But I wondered how one could learn a subject by heart without having understood it properly. Surely knowledge learnt in this fashion would be quickly forgotten and so not practically applicable? I tried to teach the students that there was an idea, a thought behind each single content. But they were not interested. I appreciated the way the German professors taught us the subject. They imparted the knowledge in such a humorous and realistic way that the material was clear, not nebulous or abstract. I loved it! When I tried to impart this feeling to others, they thought I was mad. How could I perceive it as a pleasure when it was clearly torture!

Many of the assistants and professors gave me special attention as they could tell by my questions that I was interested in the information. They took time to explain the thought processes for me. What willing helpfulness! I could have hugged them all for their effort. Many of my fellow students thought some of the assistants and professors were arrogant. But I felt on the contrary that I had the best professors and assistants in the world! When sitting in an exam and confronted with a question, which happened hundreds of times, I was interested in: 'What are they trying to find out by asking this question?' I always tried to establish a personal connection to the examiners and prepared for this in advance. My reports were always very good, often with merit. One professor was especially good to me. Thanks to his efforts I was granted Austrian citizenship, a pre-condition for being able to become an assistant.

I could enter all the rooms in the university with a master key. I could also go to the laboratories at any time and repeat any experiments until I had really understood them. If an experiment did not work, I tried it out again. Because of this I became good at operating the instruments and knew about chemical processes. Often I only went home after midnight. I finished technical chemistry with a disscrtation about a new process in producing cellulose, which was then used in the Austrian paper industry.

I loved studying and experimenting, so I had no spare time for anything else apart from music, art and hiking. I stopped doing any kind of sports during my studies. Only later I went skiing and swimming again. My two children grew up under the loving care of Gudrun. They did not have any serious illnesses or catastrophes, and had a happy childhood in a protected, orderly surrounding. My wife seldom saw me and so had to be very courageous and self sufficient. She took care of the house and school so that I could dedicate myself to my studies. One of Helmy's sayings has stayed in my memory: once, when asked who I was, he answered: 'That is the man who eats with us on Sundays!'

Despite this I remember some wonderful family holidays at the Adriatic Sea. I liked playing with my children at the beach, and found amusing things for us all to do. This lead to an occurrence which was meant to be fun but unfortunately took a dramatic turn. The children were still very young and we were playing in the sand and shallow water. I never thought about the fact that I had never learnt to swim. My mother had always warned me not to go into the water when I went rowing with my friends on the Nile because of the risk of infection. As I was playing with the children on the beach, one of my friends passed in a boat and invited us to go for a row on the sea. Mona, Helmy and I sat in the boat while he rowed far out. Then he stopped rowing and started rocking the boat back and forth for a bit of fun. The children laughed loudly. Suddenly Helmy lost his balance and fell headfirst into the water. I jumped in behind him without thinking and immediately sank like a stone. Because of the strong rocking motion Mona also fell in — all I could feel was how they grabbed on to my foot and my swimming trunks. I worked my way upwards to the surface with strong movements and tried to catch hold of the boat. My friend helped to haul the children into the boat. Once I had emerged I saw my children lying as if lifeless and we immediately started resuscitating them. I used this highly dramatic experience for us three to learn to swim as soon as possible.

I also learnt how to ski in Austria when my fellow students took me up to a hut one weekend. One day I went up to the top of the mountain with the ski lift and really let loose. I just wanted to speed down the mountainside a fast as possible in one go. While shooting down I collided with an elderly couple who were slowly meandering towards the valley. Luckily nothing happened to them. But I sprained my foot and had to be transported to the next hospital, where I was operated on my meniscus. I also started learning to glide with Helmy,

Helmy and Mona, 1966.

1. Formative Years

until Gudrun noticed and asked us to stop for safety reasons. I would rather not mention the phase I had of participating in risky car rallies on the tortuous pass roads of the Alps.

My original plan was to go back to Egypt after my graduation in technical chemistry to help my father run his business. While studying I continued following the political changes in Egypt. President Nasser was continuously at war. He sent his soldiers into Yemen and North Africa to help free the people there from their 'evil kings,' which he succeeded in doing. Africa was changed by Nasser. Due to this, many Europeans left the African continent and Egypt. Nasser wanted a socialist revolution which had drastic social results. The people of Egypt were hardly able to cope with the land reforms, the changes ensuing from building the Aswan dam and the new social conditions. My parents were also affected. My father was dispossessed and suffered major financial setbacks. He could hardly bear to see how his once blooming, laboriously established business slowly became run down by the state. For me personally this development also had

Celebrating graduation with his family, 1967.

consequences: because there were no factories left to work in and develop further, I decided to stay in Graz.

I gave chemistry lessons to students of medicine as many of them found this subject difficult. I had always wanted to study medicine myself and had only not done so because of my father, but now that I could not take over his business I signed up for medicine with the support of Professor Spitz. I assisted in biochemistry, studied medicine and helped the professors and assistants with my knowledge of chemistry. I chose to specialize in pharmacological research rather than clinical medicine. Despite this I had to take part in all the practical work experiences in the hospitals and also had to do night shifts. At the end of my pharmacological studies I graduated with a dissertation on the thyroid gland.

During my studies I saw Graz as a great university city with a college of music, a technical university and a university encompassing all faculties. Graz was a real student city. It was also a city full of culture, with an opera house, several theatres, museums and art exhibitions, philosophical circles and poetry readings. This era was one of the best times of my life. I lived in the Schiller Strasse, the continuation of which was the Goethe Strasse — what a wonderful coincidence! Of course Goethe continued to interest me, as his works had inspired me to study in Germany in the first place. But further work on Goethe's literature became something of a disappointment. I thought I would understand and enjoy this great poet better now that I could understand his language. But this was not the case. Much of what he said in his poetry and plays I could not understand at all. So I signed up for philosophy in the hope of getting to know Goethe better. This turned out to be a mistake. What saved me was the Goethe Society in Graz where the members were working on Goethe's *Faust*. The evenings started with explanatory lectures about specific scenes of the play, which helped me to understand the following recitation much better. I made my peace again with this great poet. I also got to know the works of Schiller and Herder in this society. I would often enthusiastically recite passages from them.

I was very interested in studying philosophy. Working through the development of thought from the Greeks to modern times taught me how thinking slowly evolved in the human soul. I admired the intelligent questions that philosophers like Socrates, Plato and Thomas Aquinas struggled with throughout their lives. The idea of development as such fascinated me and I enjoyed comprehensive discussions about thought with my fellow students. It was particularly enjoyable talking to one friend, who as a 'perpetual student' was seen as a failure in practical life.

During my studies in Graz I noticed other inner changes taking place. I became thoroughly involved with European culture, getting to know its music, studying its poetry and philosophy. Somebody looking into my soul would not have seen anything 'Egyptian' left, so completely had I absorbed everything new. Despite this I still felt grounded in Egyptian culture because of my upbringing during childhood and adolescence. I existed in two worlds, both of which I felt were completely different from each other: the oriental spiritual stream I was born into and the European, which I felt was my chosen course. But during this time I also started experiencing moments when these two streams met in my soul, when I was neither Egyptian nor European. This occurred particularly when I was experiencing art. For example, I started hearing Handel's 'Messiah' or Mozart's 'Requiem' with Muslim ears as praise to Allah. The two completely differing worlds within me slowly began to dissolve and merge into a third entity, so that I was neither completely the one nor the other. I could live in both worlds, think both ways. But what I experienced was not a cheap compromise, nor just tolerance, but a synthesis, even an elevation in the Goethean sense, a real uniting of the two cultures within me. This experience gave me a wonderful sense of freedom and these moments were filled with greatest happiness and joy.

Because of my family I avoided becoming too much of a European. I wanted to remain Egyptian even though I did not feel any connection to the group of Egyptians in Graz. In my eyes they had remained too Egyptian. I had become something else, and that is how I wanted to remain — even concerning my religion. I was a Muslim because of my upbringing: I did not drink alcohol, did not eat pork and continued praying regularly. But in Graz I lived in a strong Catholic community, and did not mind attending the Catholic mass where I experienced the religiousness of this belief deeply. I could live with both religions in my previously described state as existing as a 'third' entity. There were moments when everywhere I looked I could see elements in European culture which seemed like the realization of Islamic ideals. During my childhood a kind of Islamic conscience had been cultivated in me by the morality of the Koran verses, and I felt protected from adverse forces. In Europe I found that people accepted my separateness — also in my religion. Before my marriage I had told the priest that I wished to remain a Muslim. But now I wanted more — I wanted to achieve this state of being a 'third' in religion too, to be able to live within both and through this transcend to a higher level of being.

Meeting President Sadat

Shortly before the outbreak of the first Egyptian-Israel war Nasser asked his embassies to invite representative Egyptians living abroad to a conference in Alexandria. About 500 people came together from around the world at the end of the 1960's. I was chosen to represent Austria. President Nasser, his deputy Anwar el Sadat and many important ministers were the committee sitting at a long conference table. As our places were allocated according to initials, which were 'AAA' in my case, I got one of the first seats, the same as in school all those years ago. I knew Sadat personally from my adolescent years. He had been the leader of the Islamic Youth Conference, which was more or less a youth club. I played table tennis there and now and again we had lectures about Islam. Sadat was our guardian and we had met a few times. We spent a week in China as part of the Islamic Youth Conference and had spoken together on the aeroplane. In 1966 he joined the parliament and became Nasser's deputy.

In this advisory conference Nasser asked the Egyptians living abroad how Egypt should behave towards Israel. One after the other people got up and spoke out in favour of a war against Israel and expulsion of the Jews. I remained silent until Nasser asked my opinion. After a short introduction I said: 'I am in favour for peace in Israel and think even the thought of war is harmful. It can only destroy both countries and their people.' There was a great uproar in the room. Everybody talked at once and I heard words like 'Traitor!' Nasser quietened the others and asked me to continue speaking. So I told them the vision I had: If Israel and Egypt kept the peace, then the money and energy saved from supporting the war could be used for establishing a functioning economy and a cultural life for both countries.

Sadat had been observing me quietly throughout my speech. Once I had finished talking and glanced up, I met his gaze and we recognized each other again. After the conference, on the way to the photography session, Sadat took me aside, shook my hand and said: 'What you said was excellent!' and nodded at me. Later Sadat asked me if I could remain in Egypt for another week to give some speeches

1. Formative Years

at a youth meeting. We exchanged ideas and he asked me whether I belonged to a particular political party. I said no. Since then I have often wondered where this vision came from, and from whence I received the courage to speak out so openly. Sadat also asked me what I meant by establishing a cultural life. So I told him about European opera houses, universities, about museums, art and philosophy, and about the sciences and said: 'If you came to Munich or Vienna, then you could experience how beautifully the Europeans arrange their lives. But it all requires a lot of money, and we cannot afford to waste it on machine guns and weapons.' Sadat listened attentively while everyone else asked irrelevant questions or criticized me for wanting to leave Palestine to the Jews.

A year later the war broke out with Israel, and it turned into a complete catastrophe for Egypt. Thousands of people died and Egypt suffered huge land losses. This came as a great shock to many people and alienated both countries further from another. After Nasser's death in 1970 Sadat became new president as his successor and continued following his politics. In 1973 he attacked Israel together with Syria, but was forced to agree on a ceasefire after initial successes. But he succeeded in achieving what he wanted: a basis for negotiations. Gradually he eased the close relations to the USSR and started working together with western industrial states. In 1976 he abrogated the Egyptian-Soviet Treaty of Friendship. With his historic visit to Jerusalem in 1977 he initiated peace with Israel, which was sealed in 1979 with the signing of the Egyptian-Israel Treaty of Friendship despite strong opposition from Syria, Libya, Algeria, Iraq and the PLO. Because of the peace agreement the Sinai peninsula was given back to Egypt in its entirety by 1982.

In October 1981, Sadat was murdered at a military parade by fanatics who could not endure his policies of peace and cultural establishment instead of confrontation. I think back on this great politician with respect and affection. All those years I have also remained true to my vision. I still believe that war is much easier than peace. During times of peace it is necessary to work together and find each other to do well. Sadat's successor, President Mubarak, is continuing Sadat's policies and deals harshly with any kind of fanaticism or terrorism. But a lasting solution would have to include giving people education and work. Here lies a strong reason for my later return to Egypt.

Working life

After my graduation my pharmacology professor suggested I follow a university career. But belonging to the university also would have entailed other duties: joining a political party, membership in societies and attending club evenings. Political engagement was also expected, which did not appeal to me. So I left the university.
I became a research manager in the pharmaceutical industry and had the feeling I was leaving the richest and most fulfilling time of my life behind me. During the time I spent working in the industry I had to sacrifice most of my artistic and philosophic interests. The people around me were mainly interested in money. Because of this working life started off as a bitter disappointment. I often regretted my decision and would have liked to go back to university, as I noticed how my soul was wilting.

I was first employed by a medical company in Lannach, which had been established by a doctor of Jewish descent and housed in a beautiful castle. Later the company was managed by a new director. The company's licensed medicines were a success, but the development and research of new medicines had been neglected for decades. I was employed as the director of research with the task to rebuild the entire department. I was offered a very lucrative deal, which included later participation in the profits. I made plans for developing new and innovative medicines and applied for research grants. I then also administered the money the company received from the government. After three years I joined the management circle. Everything worked fine and I was able to implement many changes.

An even larger medical company in St Johann was keen to employ me, and in 1972 I moved there with my family. Again my task was to establish the research and development department, and within a short time my career blossomed. I negotiated with government bodies and research centres for funding and was able to talk about my projects in such a convincing fashion that I always received the necessary money. I had to look after state and company finances responsibly and gradually established a large team of workers. I delegated projects and sub-orders to clinics and research centres throughout the world, attended conferences in America and Japan among other places, kept contact with many German and European universities and developed new medicines, especially in the area of osteoporosis and arteriosclerosis, for which I received patents in my name.

I spent my leisure time with my family. We bought a large house with a garden in the countryside, went skiing in winter and played tennis throughout the year. I had become a very intellectual, successful, experienced and established gentleman. But I managed to avoid meetings at club evenings. Despite my middle class lifestyle, I was still interested in philosophical questions and continued educating myself; for example I read *The Story of Civilization,* the eleven volume cultural history written by Will and Ariel Durant, and studied questions concerning the development of humanity.

A lecture in St Johann

In 1972 I was asked, as an Egyptian, to give a lecture about the Israel-Egyptian conflict, which people were deeply shocked about. I was happy to oblige. During the lecture I tried to illuminate my inner thoughts on this subject, which I had also talked about at the Egyptian conference in Alexandria with Nasser and Sadat years previously. I said something like: 'Without thinking, people let themselves, their wives and their children be roused and sacrificed for emotions like national pride, dogmatisms and territorial claims. But a justification for fighting can only be seen from a higher point of view, from the ability to think about and overview complex connections. I do not believe most of my contemporaries nor politicians in the Near East have this thinking ability. The problems underlying the conflict cannot be solved by a war, only through education. People need to be educated to understand that their lives do not depend on material objects or on whether they own this or that piece of land. They need to learn to advance themselves and give their children the chance to do so too. If humans are not able to think, who is going to think for them? The devil riding them! Neither Nasser nor the Israelis are acting out of an overview of higher ideas, but out of their emotions. But people err as long as they are acting following their emotions alone. They listen to devilish inspirations, which lead them to war and destruction. If you ask me what I would do instead I would say: put all the energy, all the money into schools, into establishing the infrastructure and creating jobs. Discuss questions of cultural exchange and research and not themes that can only divide the people. I would like to shout out loud: stop, do not act until you are mature enough to be able to decide!'

I noticed a dignified old lady sitting in the front row who was listening intensely. After the lecture people stood around me for a long time, but the old lady waited until they had all left. Then she came up to me and asked if I knew about anthroposophy. I looked at her full of surprise and shook my head. She then asked whether I had ever heard of Rudolf Steiner. Here I also had to say no.

'Are you interested in finding out more?' she asked. When I said yes, she invited me to visit her at her house.

Martha Werth had two rooms in an old house, one red and one blue room, full of books, and in the centre of each room there was a grand piano. She was a piano teacher and still gave some lessons to private pupils. I was greeted by the aroma of rosemary as I entered her house. There were strange pictures on the walls which I had never seen before. She asked me to sit down, took a book off the bookshelf and showed it to me. It was called *The Philosophy of Freedom*. Did I know any works of philosophy? Of course, I had spent a lot of time studying them during my education. The she asked me to read. I opened the first page and started reading the introduction. I read her a page out loud. She listened with concentration and asked at the end of the page: 'Can you recapitulate that?'

Why not, I thought, and I told her what I believed I had read. Once I had finished she looked at me with grave astonishment and said gently: 'But, Dr Abouleish, what you just said was not in the book at all!' Now I was irritated. Actually I had given her my personal interpretation of the text and not the content of the page. But I felt her remark was a criticism and put me in my place. A mature, confident man like me could not just accept that. So I read it again, this time with full consciousness and concentration and discovered I needed huge mental effort to really understand and repeat the content. After my second attempt she was satisfied. I asked her who had written the text and she showed me a picture of Rudolf Steiner. Then she told me briefly about the book I had just read. I said a polite goodbye to this visit and thought, as I went down the stairs: 'That's it, never again!'

But the whole situation, and the text I had read, remained in my mind. The lady had given me her telephone number. Two days later I saw myself reaching for the telephone and dialling her number to ask for another appointment. She welcomed me with a huge smile. I sat on the same chair, and again she gave me *The Philosophy of Freedom* to read. She interrupted me after every paragraph and asked me to repeat the content. I let her do so as I noticed something changing within me.

After that I went to her house almost every second day. She herself did not take part in the exercise, but let me work alone, reading the content and then repeating it in my own words. She described briefly how this exercise fitted in with the complete works of the author. For me, the crucial biographical experience lay in the fact I began to experience the act of thinking itself through the enormous mental effort. This in turn led to the ability to handle and organize the things around me in a much more attentive and efficient way. This effort transformed me internally and externally, and I surrendered myself to what was happening, which was like a resurrection, like a remembrance of things long known, filling me with enthusiasm. All my solely intellectual and clever knowledge slowly began to be transformed and I could see it in a different light, not so much because of the content of what I read, but because of the quality of the mental effort involved. Thus I developed a deep love towards this anthroposophy. I had the feeling that through it I had grasped a tiny part of the whole world, and humans and nature were revealed to me in a new light.

The old lady and I grew very close. After some time I asked her to give piano lessons to our children Helmy and Mona. Helmy was fifteen years old at the time and asked her once what his father did during all the hours he spent visiting her. When she told him, he also wanted to read this book. As he continued persistently asking her, she started working with him after the piano lessons.

While gaining more insight into the works of Rudolf Steiner, I came across his talks about the Old and New Testament. I read these spiritual descriptions as a Muslim, and was impressed not so much by the content but by the way these themes were illuminated and deepened. I wanted to work through the Koran using anthroposophy to achieve a new, deeper understanding. This was the seed which led to my much later efforts of interpreting the Koran in a spiritually deeper way. What sounds so easy in retrospect had to be attained gradually with intense internal struggle, a daily observation of my relationship with Christianity and European culture.

After we finished *The Philosophy of Freedom,* Martha Werth studied further works of this author with me, working through them in the same way by letting me first read and then recapitulate the content. My philosophical interests were given new nourishment and impulse. But even more occurred: because of the intense mental effort involved in this work an unnoticed change took place in my soul. In retrospect this is how I explain the effect the journey to Egypt had on me which my family and her were about to undertake.

Opposite: Martha Werth and Ibrahim Abouleish in Upper Egypt in 1975.

2. Return to Egypt

A journey

'Wouldn't you like to join me on a journey to Egypt?' Martha Werth asked us one day. She wanted to know if I had already come across ancient Egyptian culture. I had actually lived near Gizeh the year before my final school-leaving exams and had a panoramic view of the pyramids from my window. I greatly valued the treasures of ancient Egypt. Now Martha Werth was asking us to accompany her on a trip around Egypt, and we started preparing for this great event.

While living in Europe I had visited Egypt repeatedly at my mother's request. But I had never experienced such an intense encounter with my culture as on the journey with Martha Werth, which became a turning point in my life.

We started our journey in 1975 and visited the many famous ancient Egyptian sites in Aswan, Luxor, Karnak and the Valley of the Kings. I can still see Martha Werth today, striding ahead of us with energetic steps and a red sun umbrella, a travel guide clutched under her arm. My understanding of the buildings and artworks was deepened by her explanations, and I began to see them in a new light. She gave me new enthusiasm for ancient Egyptian art and mythology.

In Luxor we took the ferry from our hotel across the Nile to the western bank and then travelled to the temples of Hatshepsut by taxi. Martha Werth felt all purely material and historical explanations of these cultural artefacts were imposed from the outside and instead tried to find an individual, authentic approach to understanding the works of art. She pointed out how the temple and its pillars fitted into the surrounding landscape, and explained their relationship to the vertical rock formations of the western mountains, El Qorna, in the background. The architectonic artwork nestles in amazing natural surroundings. I realized how nature is transformed and elevated by the artistic creation of the temple's buildings. We alighted beneath the broad entrance and slowly approached the wide forecourt and high columned hall of the temple. Martha Werth stopped in awe to feel the fine, precise contours of the wonderful reliefs decorating the pillars and walls of the whole temple. We followed her example and I was amazed by the artistic skill of the millennium old stonemasonry.

In the Valley of the Kings we descended into Tutankhamen's tomb where the rich gold artworks were discovered. Later we admired the treasure in all its glory in the National Museum of Cairo. The fine goldsmith work and the use of the jewels lapis lazuli and carnelian together with gold deeply impressed me. The three primary colours blue, red and yellow harmonized beautifully in these wonderful artworks, not to mention the precision and fineness of their making.

In the other tombs of the Valley of the Kings I was amazed by the modernity of the depictions of the soul's travels after death, which covered the walls in abstract sketches. Martha Werth explained the significance and background of the mortuary barges, the symbols and the encounters of the king with the gods. The boundaries of an earthly life became permeable and expanded into other dimensions.

In the temples of Karnak I was particularly impressed by the two obelisks of King Thutmosis I and Queen Hatshepsut in the temple

2. Return to Egypt

of the God Ammon. My eyes followed the tapered, exact lines that ended in a triangle pointing heavenwards. I felt as if I was an imitation of this symbol, as if internally aligned, and thought of a similar experience which I had had as a young man coming to Austria, when I first saw and admired the wonderfully tall trees, the beech and fir crowns of Europe's trees. The people of Egypt do not have the chance to experience the humanizing quality radiating from tall trees. Were the pharaohs aware of this and to compensate erected the obelisks for schooling inner uprightness? Again we admired the play of light and shade created by the three-dimensional symbols and figures on the pillars.

But it was not only ancient Egypt which was revealed to me in a new light. Through visiting friends and relatives, and particularly a journalist acquaintance, I became immersed in everyday Egyptian life. Partially due to the amount of time which had elapsed since I left the country at the age of nineteen, I noticed the great changes which had befallen Egypt whilst I had been in Austria. My friend introduced me to several politicians, and we discussed the changes that I had observed during the previous weeks. One of the politicians said: 'Egypt used to be a wealthy country during the 1920's and 1930's. The Egyptian pound was as strong as the British pound. Although rich and poor existed the rich looked after the poor. Consideration for others, courageousness and a deeply moral attitude towards humans and animals were typical qualities of the Egyptian people. Because of the low population of only eighteen million people, Egypt was a beautiful country and Cairo a thriving city.'

'I remember the streets were cleaned daily,' another one added. 'Everything was tidy and looked after, mainly due to the Europeans, as Cairo was a multicultural city with many European businesses.'

'The only unpleasant thing in our eyes was the unfair treatment by the English. But at least we could rule ourselves through a multi-party parliament,' my friend said.

'That was twenty years ago,' I said. 'During Nasser's short time of presidency Egypt has changed completely. The once healthy, cheerful population seems to have sunk into a deep depression. The cities are dirty and there are terrible rubbish dumps everywhere ...'

'Yes, our health service is miserable and too little is spent on treating illnesses, so the people are wasting away,' my friend added. 'On top of that we have caught completely new illnesses, which I wouldn't have thought possible in this country where the people ate healthily, did not have any stress and did not smoke. Now the opposite is true! And we also have stomach, intestinal and parasitic illnesses because of the lack of hygiene.' These discussions with the politicians painted a dismal picture of Egypt.

Under the reign of Nasser all businesses had been nationalized, even the restaurants. Once thriving businesses were now working with deficits, most people had jobs they did not enjoy doing and worked without inner motivation, and many people were forced to take jobs on the side. The whole social structure was increasingly falling apart, leading to immense misery.

I experienced agriculture as a catastrophe. The farmers were forced to use a certain amount of artificial fertilizer per hectare of land. This excessive and uncontrolled use of fertilizer led to oversalting and compression of the earth, not to mention the financial dependency of the farmers who had to buy the products. The inheritance laws of the country assigned equal land division to the inheritors, leading to smaller and smaller plots of land with each generation. The farmers could hardly produce enough to survive. Added to that was the appalling spraying of pesticides onto the cotton fields. The Aswan dam, completed in 1961 with the Soviet Union's support, also had disastrous results for agriculture. Since then the Nile, which had previously flooded its banks every summer and spread fertile mud over the fields, had ceased to be the pulsating heart of Egypt. A year round irrigation system led to standing water in canals becoming a hotbed for dangerous diseases. The hope of gaining more fertile land through this irrigation system was not fulfilled. Naturally the dam made it possible to produce electricity, but this electricity was mainly used to manufacture costly artificial fertilizers.

I travelled around the country with the journalist and looked at the schools. There, too, I was presented with a hopeless picture. There were not enough teachers and the classes were completely overfilled with up to seventy pupils in each class. The state did not concern itself with the resulting problems as it was preoccupied with the war. Illiteracy was on the increase. More than forty percent of children did not go to school at all because they had to work to help support their families. Those who did go to school learnt the material strictly by rote, rather than by creative, artistic methods — a pedagogy which had to lead to further misery.

'I can predict what will happen if things continue like this,' I said to my friend the journalist. 'The people will leave their farms and go to

2. Return to Egypt

the cities, where they will be left to fend for themselves. Where else can they go to live and work? Even now they have started settling in the city of the dead, the old graveyard of Cairo. Isn't that a terrible and unworthy place to live? The worst thing about the cities is all the rubbish and the expanding slums!' My friend confirmed that even educated people did not know how the country should continue. He felt the government had already failed.

My visits to the mosques revealed further problems. The oldest university in Egypt is in Cairo: the Al Azhar University, where Islamic religious-philosophical questions have been researched for centuries. Countless books have been written, long, detailed articles have appeared in magazines, never-ending discussions have been held. But I could not see the results of these intellectual efforts in practical life, in the deeds of our modern twentieth century times. Islam has a very conservative view of everyday life. Things the Prophet said in the seventh century are still used as a basis for acting in many areas of life, even though in those days industry and economics did not exist. Laws and jurisdiction are partly adopted straight from Europe or, as in the Sharia, handed down rather than developed specifically for the Islamic population using modern cultural circumstances. Discussions conducted with Egyptian scientists about possible changes remained on a superficial level.

I experienced the people themselves as very religious. Many of them kept to the praying times and visited the mosques regularly. Translated, Islam means 'submission to God,' and this attitude is deeply embedded in the Islamic-Muslim tradition: living in intimacy with God, turning to Allah, the omnipotent, for everything. But at the same time I was shocked by the discrepancy between the inner, religious human who would never harm the ground, plants, animals or fellow humans and the person acting in the working world. In everyday life I experienced Muslims working for egoistic, compulsive reasons, not out of a deep connection to religion. The earth and money are treated as if they belonged to the individual for all eternity. They attach themselves to material things, and even treat their children as if they possessed them. They often appeared dogmatic, making great speeches, but seldom following them up with the right kind of deeds. The deeply felt religion was usually limited to a personal, private area and did not radiate into practical life.

Because of my observations, the lack of innovation in the legal and economic sphere and a religiosity limited to personal life, I received the impression that Islam and its people had stagnated and were in a deep crisis. I was shocked by the contrast between the greatness, wisdom and elevation shown thousands of years ago by the pharaohs, and modern Egypt. In the evenings, after visiting museums, I discussed my experiences with Martha Werth. She felt my disturbance, listened attentively to my questions and then said: 'What do you want to do? Its destiny! Where can one begin?' I kept comparing my memory of the country of my childhood and adolescence with the country I found now. With shock I realized that the old one was often better than the new — but in a life orientated towards the future the new should have been better than the old. Here it was different — such a decline in twenty years!

My heart tries to understand

On my return journey I sat in the plane and thanked Allah that I did not live in Egypt, but in beautiful Austria with my wife and two children and my successful career. And yet I could not forget the images and encounters I had experienced. Every morning I awoke and realized anew how the events of the journey had transformed me. I obtained further information about conditions in Egypt. During my trip I had become aware of the excessive use of artificial fertilizer and pesticides and the resulting oversalting of the earth. Now, in Austria, I discovered far worse things about Egypt's economy, education and health situation and agriculture and trade relations than I had already learnt through my discussions with Egyptians. I talked about these things to the Egyptian ambassadors in Bonn and Vienna, although I had the impression that they tended to make the situation appear better than it actually was. Once I showed them the facts and figures which I had collected for an extensive study they were very shocked. I did not want to blame anyone; I wanted to discuss the problems with other people to receive more information. I then tried to deal with the resulting feelings on my own.
At the same time I continued to work with anthroposophy and became acquainted with its practical applications in many areas of life. The deeper I was able to penetrate into the matter, the more answers I received for my persistent questioning and inner restlessness. I repeatedly found life changing solutions suddenly presenting themselves to me after intense contemplation. Biodynamic agriculture particularly

2. Return to Egypt

Georg Merckens, 1990.

fascinated me. It was developed out of anthroposophy, the philosophy developed by Rudolf Steiner, and has been practised successfully in Europe since the beginning of the twentieth century. I was sure that it could improve the agricultural situation in Egypt. But many questions remained unanswered, so I went to ask Martha Werth about them. One day she told me about a lecture by Georg Merckens in St Johann. Merckens was the advisor of the biodynamic farms in Austria and Italy. He was a magical storyteller, who lectured with a wonderful voice and great clarity. After the lecture I talked to him to find out more about biodynamic farming. We arranged a week were I could accompany him on a tour throughout Italy. During the journey he would tell me about biodynamic farming and at the same time show me practical examples. At last I felt I had found a friend who understood my idea that biodynamic farming could transform Egypt's agriculture. In my opinion one had to start by establishing a self-sustaining farm, and then add more projects. Georg Merckens was willing to partake in my plan, although he kept asking with surprise: 'Where do you receive the courage to do this as a non-farmer?'

In this phase of my development, my urge for knowledge knew no bounds. I felt there was nothing I could not know, nothing I could not do. Where were my limits? I had not come across them. During my life I had acquired three abilities which I could trust: firstly, a great ability to learn. Secondly, I experienced time and again that I could meet people and win them over, and lastly I had tremendous creative energy. I thought that I would be able to do anything if I only had enough time to plan and prepare. I had to admit that time really was my limitation. But I learnt to set priorities, and to know what was most important for me at any given time. Now I had to give precedence to learning about biodynamic farming. This I did while travelling through Italy with Georg Merckens. During the long drives from farm to farm, I acted as Georg Mercken's chauffeur; he explained the basic ideas of biodynamic cultivation. We were always welcomed warmly by the farmers of the large Italian farms. I absorbed everything we discussed and I observed about agriculture. While striding through the fields or stables he showed me practical examples of what he felt had worked well biodynamically and what had not worked so well. Every evening there was a meeting with the farmers where he gave background information and answered questions. I listened with great concentration to all this, had interesting discussions with the farmers and believed I had soon found the critical weak point of biodynamic farming: it was a lack of knowledge about how to market the products.

Georg Merckens, too, was not a marketing expert. When I asked him about the principle of 'association' suggested by Steiner — a relationship between all people involved in the economic process leading to a mutual understanding and fulfilment of the needs of everyone — he brushed it aside and said: 'We are still miles away from being that far. It is still an ideal. We can't do that yet!' But I was convinced that in Egypt this point would have to be tackled very consciously right from the start.

At the end of the week I drove back to Ulm with him. In Bad Waldsee he introduced me to Roland Schaette. 'If you want to establish a farm, you need to get to know a business which produces organic animal fodder, veterinary medicines and pest control agents.' The young scientist had just graduated from Professor Wagner in

Munich. Enthusiastically he told me about his work with valerian, gave me his dissertation to read and showed me around his business. His company seemed small and unpretentious and was equipped with simple apparatus. But it had more than fifty years' experience in the area of organic veterinary medicines. Roland Schaette met me with such openness that I became very fond of him. He was also one of the few people who took my questions seriously and did not write me off as an amateur. He was willing to listen to everything I had to say and I took leave of him in the hope that we would be able to work together.

My 'Italian journey' was an important step along the path towards my decision to return to Egypt. I developed a vision of a holistic project able to bring about a cultural renewal. As well as the farm it would need one or several economic projects, a school and different educational institutions and offer cultural projects and medical care. My first priority was to educate people. But I would need to create concrete institutions for all this so that the project did not remain solely an ideal. So I started to look for people to work with. I knew I wanted to implement an independent project without the help of state funds.

I also knew I would encounter difficulties with the Egyptian authorities (they turned out to be a lot worse than I had ever anticipated). I hoped to find idealistic people who would be inspired by a new cultural project. In the nineteenth century, Mohammed Ali had accomplished a similar task when he called upon Europeans to help rebuild Egypt. It would not succeed with the help of Egyptians alone. But I was certain that a cultural meeting between Egyptians and Europeans could become a healing force in this oppressed country.

I discussed the idea with several Egyptian doctors and farmers I knew. They were delighted with the idea, but did not think it could work in reality. Some of them told me they had tried to change things in Egypt, but had failed because of Egyptian bureaucracy. They advised me against trying to do anything in that line. I encountered this attitude repeatedly when I talked to people about my idea — they could not really understand it with their heart, or to put it differently, that they did not possess the courage to act.

I visited the Egyptian ambassadress in Bonn, an intelligent woman, and discussed my ideas which were slowly crystallizing into concrete plans. She said: 'That is an amazing dream you are talking about. Wonderful!' — 'Tell the government that something like that can take place!' I answered. But nothing happened. I could not find anyone who was willing to partake in the project.

Three years had passed since my last journey to Egypt, and my thoughts of returning had taken shape during this time. I would have found it unbearable to give up just because I could not find anyone willing to join me. So I decided to go it alone.

Farewell to Europe

How did my family react to my decision to move to Egypt? My wife Gudrun loved Egypt, and particularly my father, and this strong inner reason sufficed for her to want to join me. I told our children the story of a man who decided to move to the desert with his children and who created a big garden there. Once I had painted the picture in great detail, I suddenly asked: 'And what would happen if we were that family?' Spontaneous shouts of joy followed the question. My son Helmy was sixteen years old at the time and my mother had already told him about all the things I had done at his age in Egypt and which were not possible in Austria — driving a motorbike in the desert for example. He wanted to get to know the oases and mountains of the Sinai — a wonderful reason! And my daughter Mona, fourteen years old, was in love with horses. She would be able to ride as long as she wanted to in the desert. Thus everyone was inspired by their own inner vision.

And then the madness started: we lived in a large house in St Johann that could not be sold so quickly. I bought three VW buses for the move and filled them mainly with books and clothes. Helmy, who did not have a driving licence but was an excellent driver as I had taught him from an early age, drove one of the busses. We drove one behind each other, me at the front, Helmy in the middle and Gudrun in the last bus, across the Alpine passes between Austria and Italy to Venice. At the Italian border Helmy, adorned with sunglasses and a hat, was accepted as a man and waved through. So the 'holy family' moved to Egypt!

I tried to outline my reasons for returning to Egypt in two farewell letters to friends in Vienna. To Martha Werth I wrote:

> *Austria is a paradise. The mountains, forests, meadows and lakes are like a huge garden, a wonderful park. People have created all this over*

2. Return to Egypt

the centuries with loving care and through it have elevated nature. Egypt does not have all this. I am consciously leaving the Austrian natural beauties, which gives me so much energy, to go the desert and create a garden with forests and tree-lined avenues, roses, fruit trees, vineyards, meadows, fields with aromatic herbs and animals. I hope people will be able to experience what I received from this wonderful Austrian landscape: because of nature's beauty I experienced the inner light of thinking radiating within me. For my soul Austria was like a spiritual childhood garden. Now I hope that the souls of Egyptian people can be revitalized by a garden in the desert. After establishing a farm as a healthy physical basis for soul and spiritual development, I will set up further things, following the example of human development: a kindergarten, a school, a vocational school, a hospital and various cultural institutions. My goal is the development of humans in a comprehensive sense — educating children and adults, teachers, doctors and farmers. Not only did I receive so much from the landscape of Austria, but the people there met me with liberality and generosity. I took with a full heart what was offered and deeply absorbed the culture. I want to pass on this richness of nature and spirit to Egypt, to sow the seeds I have been given.

I wrote a further letter to my friend Dr Johannes Zwieauer in Vienna, who according to Martha Werth was an authority who should have dissuaded me from my decision:

I have decided to leave Austria to start a farm in the desert in Egypt based on a holistic developmental impulse for country and people. Partially I see the reason for this decision coming from my occupation with anthroposophy. It has deeply influenced me.

My soul has begun to separate into two parts: an ambitious successful part and a seeking questioning part, willing to see things in a new light, and to transform and elevate them to a higher level. I am consciously leaving the successful part behind me and am giving myself up to the questioning one. With this I am uniting my soul with its spiritual home and am liberating the rigidity of ambitiousness so that I am open for new tasks, encounters and goals.

Ibrahim Abouleish in 1977 just before returning to Egypt.

Returning home

My heart broke in two on the ferry. One side of my heart shed real tears. I was in the process of giving up a successful career as a researcher to exchange it for an incredibly unpredictable future. But I was also consciously leaving a part of me behind. Twenty-one years before, as I was departing from Alexandria on a Turkish ship to immigrate to Austria, I said farewell to my mother and my native country. Now I was sorrowing deeply about the loss of my chosen spiritual country. How I would miss the operas of Salzburg and Bayreuth, the *Faust* plays, the discussions with friends and the philosophic readings! I felt deep loneliness and had the distinct feeling that it would take some time to feel at home in Egypt again. But I could soothe my heart with the words of Hermann Hesse: '... A magic dwells in each beginning, protecting us and helping us to live!'

2. Return to Egypt

The other side of my heart was deeply aware of the three abilities waiting to be sown like seeds in Egypt's soil, to germinate, grow and shape anew. On my last journey through Egypt I had experienced a deep sense of hopelessness caused by the way of life of the Egyptian population. This had deeply moved me, as I knew that people's surroundings mirror their soul's disposition. I felt compassion for these people who could not be made responsible for their situation, but were forced to bear it and had learnt to carry it. My work with anthroposophy led me to sense a way that could liberate them from their misery.

I have already mentioned the three abilities that had grown within me throughout my life, wherever they may have come from: the ability to learn, social skills and my energy for doing things. Because of these abilities I felt I would be able change this situation of hopelessness. I felt privileged that I was allowed to take on the task of establishing something in the desert! It is truly lucky if one feels equipped with the right 'tools' for one's task. With this feeling of happiness I hoped to find people who followed the same ideals and would support me with their abilities and energy.

My faith in God gave me an inner strength which had grown out of years of meditation on Allah's qualities in particular. I asked myself what the Koran meant by stating: 'He is the representative.' During my journey through Egypt I noticed desolation everywhere in the population, a desolation which these people did not even feel themselves. But I felt this physical and spiritual-soul emptiness in their stead, and thus I experienced myself as their representative. Because of this awareness I wanted to establish new social forms for the Egyptian people.

The Koran goes on to say: 'He is the initiator, the originator, the strong one.' Was it not a privilege to be able to act? Nowadays all paths seem to be well-trodden. Where is it usually possible to initiate something new?

'He is the powerful one,' the Koran says. I felt power in me for this new start as I wanted to achieve it with consciousness.

I was able to develop inner peace through my devotion to Allah, and even today I can still submerge myself in its depths.

Part 2

Founding a Desert Community

Opposite: The road to the farm in 1978.

3. The Beginning

Desert country

After arriving in Egypt I first went to visit the minister of agriculture. I explained to him that I was looking for a patch of desert, which I wanted to cultivate using organic methods. It was a sign of his friendliness that this busy man listened to me for half an hour. After our conversation he asked an employee of the ministry of agriculture to show me some areas of desert I could buy off the state. After all, there was enough desert in Egypt. 'It will be easy to find desert!' said Kamel Zahran, an old, honourable, high-ranking engineer. First we drove west towards Alexandria. From the asphalt road he pointed out areas of land for sale which had good access to water. The minister said he could put in a good word for me if I wanted to buy the land. I looked at everything, asked about the people living there, about possible energy sources and whether roads could be built. But inside I remained untouched. This happened on the first day, as well as the second day.

On the morning of the third day my contact said he had to visit someone before we continued our trip, as he was also an agricultural councillor. He needed to visit a farm north-east of Cairo, at the Ismailia Canal. Because I was his chauffeur he asked me to drive him there. We left the car at the canal, took the ferry over the water and arrived at the farm, which was a large orange plantation. My companion introduced me and explained my intent, to which the farmer said smiling, after spreading out his arms into the landscape: 'You will be sure to find something here!' After Zahran had finished his visit, we walked across the plot of land, a strip which reached about four kilometres into the desert, as far as the canal's water could reach. It was a hot day and the old man was suffering and walked with difficulty through the rows of plantation trees. Sweat poured down his face. At the edge of the estate we stood and looked out over the stony wasteland. Kamel Zahran said: 'It is impossible here. We are four kilometres away from the canal and the desert is still going uphill. We are probably already thirty metres up. You will never get the water to reach this far.' While he waited in the shade of a tree, I walked on by myself. The country, which stretched out barren and empty towards the horizon, was gently hilly. I liked the fact that it was not as flat as the delta. After a few more steps in the shimmering heat I noticed how a vision appeared before my inner eye: wells, trees, green plants and fragrant flower, animals, compost heaps, houses and working people. I would have to expend a lot of energy to cultivate such an impassable, difficult surrounding and to transform this wasteland into a garden! But many jobs could be created in doing so, and these people would have the chance to educate themselves while creating something healing for the landscape!

I walked back to Kamel Zahran deep in thought, and was immediately greeted with the words: 'It's too steep, you could never cultivate here.' But I felt I had been touched by this land, as if something had spoken to me. When I look back I have to admit my immense naivety, as I did not have the faintest idea what it meant to cultivate and irrigate land in the desert.

3. The Beginning

On the return journey I spoke to Kamel Zahran. 'You know,' he said, 'Let's not rush anything! We'll come back later with specialists who can advise us.' So we returned But after only a short time the specialist delivered the discouraging verdict: the quality of the ground was very bad and water supply difficult; there was no direct road to Cairo and all products would have to be transported via the ferry on the Ismailia Canal. The general opinion was that the land was not suitable.

But overnight I reached a decision — and by the next morning I knew I wanted to buy this piece of land. If biodynamic farming and everything else I envisaged could thrive in this wasteland and under such extremely adverse conditions, then it would be possible to transfer this model to easier environments and we would develop immense energy through overcoming such difficulties!

As soon as I had signed the bill of sale the first difficulties started. When I tried to get the plans of the seventy hectares of land I had bought to mark out the boundaries, I was told that although the state administered the land, it could not find out about it that easily. There were no surveying points and I soon noticed that the Egyptian land surveyors responsible for this area had difficulties dealing with plans and committing themselves. In those days it took three hours to get to Cairo from the Ismailia Canal by car, and I had to regard it as a favour if the surveyors even managed to arrive at my plot of wasteland, even though they were paid for their effort. When I asked Kamel Zahran for advice regarding this, he only said with *Schadenfreude*: 'Didn't I tell you it wouldn't work?' But I was not put off by all this. Quite the opposite: it made it all the more attractive and strengthened me in my resolve.

After buying the land a period of intense planning began. I tried to survey the 700 × 1000 metres myself by borrowing the necessary equipment. I struck iron poles into the sand at specific spots, and carefully drew everything onto paper. For ten years, I only had a vague idea of the boundaries, although later corrections were surprisingly minor.

First I marked out the roads: I wanted a main road to go right through the middle of the grounds lengthways from north-west to south-east. I then planned further roads branching off at right angles to the right and left of it, dividing the land into about three hectare plots for fields. In my mind's eye the roads were lined with shade-giving trees. I wanted a thirty metre wide band of trees to encircle the entire grounds, to protect the developing life of the plants, animals and humans. I used the image of a cell for inspiration as it is surrounded by a membrane. What the clear blue sky and warmth-giving sun means for a European is a shade-giving tree for the desert people. They like to spend time in the coolness of shade, and at the same time they protect themselves from excessively strong cosmic influences.

Water is of prime importance for life to be able to flourish in the desert. I decided to bore wells, one in the north-west near where I wanted to build the stables, and the second one in the south-east nearby the planned houses and living quarters. I left a long strip of land in the west for a school, a medical centre and an institution for eurythmy, art and social activities. Right in the middle of the grounds I left a space for the businesses. It was my intent that their profits would finance the establishment and development of the cultural institutions. I drew round flowerbeds on to distinctive crossings of the right-angled roads to artistically fashion the desert from the onset.

The first living house we built was round, and to start with we also made the company buildings with many curves. This impulse came from a social feeling deeply rooted in my soul: the meaning of the sun's shape for the cosmos is transformed on earth as an inclination towards social forms. I did not think at length about this image, I felt it as an inner necessity from the depths of my soul.

This first plan still exists. When I look at it today, I can see myself striding alone over the stony bleak ground, sketching and planning, unprotected from the sun and wind. This was in 1977. The infrastructure, which the visitor finds in place today — asphalt roads, electricity and telephone lines and even the villages — have only gradually developed with Sekem's development. In this same place a flourishing oasis has grown. It is almost an unbelievable miracle that the one has developed out of the other. People came, helped, stayed or left again; we had to struggle with intolerance, negligence and many hindrances. In retrospect it all appears like a living fabric, Sekem as it is today has been woven out of many strands of warp and weft. My planning bureau still exists in the top story of the art house.

The next thing I did was to buy a tractor and start building the roads according to the plans I had drawn. Most of the time I was all alone, only now and again Bedouin would wander past with their goats. They seemed to think what I was doing was very funny, looking at a piece of paper every now and again and then working. They could not understand my idea. But they saw it develop before their eyes.

The Bedouin, who were obviously attracted to just this piece of land, were friendly. I repeatedly brought them things back from Cairo, and gradually I became acquainted with them. I was touched

3. The Beginning

and thought it was wonderful that in that first year about forty people came to live and build their straw huts on my land. I felt they were sent from heaven, and started giving some of them small tasks to do. Once I was deeply engrossed in my work when someone patted me on the shoulder and said: 'I am Mohammed, your watchman!' He was a Bedouin with frighteningly ugly teeth, but extraordinary open and kind; he is still living in Sekem today. In those days I thought he had made his offer out of pure friendliness and was glad to accept. Only later I also understood his proposition it in a different way.

In the year 1978 we had a hard winter. I needed a fur hat and a thick wool coat to stay warm and it made me cold just seeing the Bedouin with their thin clothing. So I bought wool blankets for my watchman Mohammed and his family, for which they were very grateful. But one day, when springtime came and it got warm again, I was shocked to see what Mohammed was doing. He had made the wool blankets into a rope and was pulling his tractor with it. The blankets were torn and dirty. I said: 'Mohammed, what have you done? What are you doing with those blankets?' He looked at me with surprise and as if

On the tractor in 1977.

I came from a different planet. 'Mohammed, winter will come again. You will need the blankets again!' — 'What do you mean? Winter will come again? When the winter comes, Allah will tell you to buy me new blankets!'

I slowly learnt that many of the people I dealt with had no concept of time, or put differently, experienced time in a different way. But it is impossible to plan ahead, set goals, analyse, correct oneself or reflect on one's actions with this concept of time. At the same time I saw the amazing warmheartedness and openness of these people who lived completely in the moment in their feelings, who dealt out of their current state of mind. But I wanted to develop something and follow goals, for which I had to plan in advance. All my experiences showed me the importance of moral deeds as an example for these people living primarily in their feelings. The Prophet says every one of you is a shepherd, and everyone is responsible for those under your protection. By following a concrete example, which the people here imitate right down to posture and movement, they can transform themselves and their surroundings. So I would take a rake and practice raking the paths with the people, or paint the walls a shining white with them. It often sufficed for these simple people to see that I thought these things were important, and then they would try hard just because they knew I cared so much. Over a longer period of time this imitating behaviour leads to changes within the people.

At the same time I felt it was important to awaken people's wonder by exposing them to art and science, which leads them to independent questioning. When questions arise from the inside it shows the beginning of an understanding of the internal working of things. The next step, especially for these people living in their feelings, is to establish concrete social forms. This involves forms which are taken for granted by most Europeans, but are of utmost importance for the functioning of an establishment: How and when do we start our day? How do we stand in a circle, how do we dress for work? How do we treat each other in a dignified manner? Outer forms help towards developing the mind. This starts with very elementary examples: to start work punctually at seven you need many preparatory considerations: getting up on time, getting dressed and catching the bus. This forces the people to think about things they would never have otherwise dealt with. In most cases Egyptians start working when they have slept sufficiently and finish when they get tired.

3. The Beginning

Water

After I had positioned the first roads and plotted the fields, the next task was to drill two wells. I myself did not know how to a build a well, and so for the first time I was in the lucky position of needing to employ people. I heard about a team of people who were able to build a well in the desert. By hand, they made thousands of bricks in wooden models with the cement and sand I brought them. All the necessary water had to be transported in a tank by tractor from the canal. By now the surrounding inhabitants had become aware of what I was doing and watched these transports with suspicion. What is happening out there in the desert? They felt increasingly disturbed and began to defend themselves by blocking the road with stones.

'Where do you want your well?' the team leader asked. 'There!' and I drew a two to three metre circumference circle in the sand with a stick. With their hands and simple spades, the men started digging up the first half metre of sand and then built a circular wall with the cement bricks into the hollow. Overnight the cement hardened. Early the next morning they continued gradually deepening the hole, digging and hacking with simple tools even when they hit rocks. Then they built a second circular wall underneath the first. They fastened evenly spaced iron rods into the bricks as steps, so that rubble and clumps of earth and sand could be hauled out of the depths with buckets. Suddenly, after 25 metres, the sand started getting wet. They continued deepening the well another five metres, and then the well digger came. He fastened a three metre long iron pipe with a circumference of thirty centimetres to a winch at the top of the well. The pipe had a rubber valve at the bottom, and when it rushed down into the depths this valve opened and let the fine sand into the pipe. Then it was drawn up, emptied and let down again. At the same time a second pipe was let down with the first. This meant there was a pipe within a pipe; the outer one had a copper wire filter full of holes. For three months many men worked with the iron pipe down to a depth of seventy or eighty metres. Towards the end lots of water was transported up with the sand — the first water! It was tried enthusiastically by all. The first water on our own grounds — what a moment of rejoicing! But it was deep, deep down in the earth. How should it be raised to the surface?

During the day I supervised the building of the well to prevent accidents happening and to ensure the work was done properly. Late at night, once the men had left, I studied books about pumps and calculated the extraction capacity of the equipment. Once I had worked out what I needed I started looking for the equipment in Cairo, but it turned out you could not get it there. Finally I found the scrap dealers of Cairo had what I required, although they themselves had no idea what they were offering: an old electric engine, rotary pumps and check valves. Naturally I could have modern machines sent to me from Europe. But I admit I wanted to try and get all the necessary things from this country for establishing my idea for this country without any foreign help. It was all a tremendous effort, because if the machines did not work after I had tried them out I had to travel back to Cairo for three hours over the dirt roads to exchange them or find new ones.

The pumps and all other electric appliances needed electricity — but where could I get electricity in the middle of the desert? I obtained a diesel generator for which I then needed a diesel tank and a steady supply of diesel. I would prefer not to mention the many difficulties and tragedies involved in this process. Finally we were able to draw the water from the well! We were all extremely happy with these precious drops of water coming from the depths, extracted with such difficulty! We celebrated the occasion with a big party!

We did not repeat this method of building a well, already used by the ancient Romans. In total we dug five wells on the farm, with a depth of 100–110 metres. After our first experience we used a borehole pump with an electric cable to dig them; although it made the process more expensive it was also much simpler and quicker.

During this work my thoughts and plans were already geared towards the next step. I realized that because of the difficulties with energy supply and obtaining spare parts and the extreme adversities of the desert I would need to ensure that we always had spare machines, in case one of the pumps or a part of a pump should break, as the entire running of the farm depended on these machines working! For this reason I constructed a reserve system for the pumps, which started up as soon as one of the machines failed, a frequent occurrence. One time I was installing an electric pump 25 metres down in the well. After twelve hours of work I was exhausted, but had managed to get it running. At my side was a young man, Hammad, who had observed how hard I worked. I showed him how to connect the replacement machine but told him not to continue working without me. Then I drove to Cairo to sleep. Now my good friend wanted to do me a favour and show me what he had learned. Carefully he took the second pump to the edge of the well. There it slipped out of his hands and fell onto the first, working pump, and everything broke! The next morning he stood at my doorstep in Cairo in tears and I had to calm him — and start anew. I said something like: 'Listen! Just as we can learn from our achievements we have to realize our failures are also trying to tell us something. It is not bad to make mistakes. But you have to learn from them. Where did it go wrong?' As a pedagogue, I was always trying to encourage people to think about what they were doing, so they could correct themselves and not just ascribe everything to Allah's will. This is an arduous but beneficial endeavour, as Egyptians not only have difficulties thinking and planning ahead, but also reflecting on things and self-control. Hammad learnt a lot from his shock and is still with us today.

Even before the first water was pumped up from the depths I had been thinking about its distribution, or better said: planned the irrigation network. How should we channel the water so that it could reach the plants and animals? — You need a clever plan to irrigate effectively. Canals have to be built and pipes laid. Nowadays Sekem is criss-crossed by a huge underground irrigation system. But in those days I tackled the problem by purchasing a machine from Germany called a Waldhauser. This was a 300 metre long, thick hose rolled up on a reel with a water jet at the front. The hose could be pulled out over the field, whereby the water pressure caused the machine to spray water out of the jet at the front and at the same time to slowly roll the hose back up onto the reel. Then the hose was transported to the next site which needed watering by tractor. We laid thick cement pipes underground, able to withstand a high pressure. The entire stony ground had to be dug up for this. Every 100–200 metres we built taps protruding out of the ground for the machines to be attached to. As the water did not come out of the wells with sufficient pressure to run the machine, we built a pumping station to add additional pressure, which was then passed through the pipes. Everything had to be coordinated together: the thickness of the pipes, the pressure created by the pumping station, and the water density. I had to learn everything carefully step by step, as even the cement pipes could explode if the water pressure was too high.

After we had installed all the necessary prerequisites I realized that this irrigation system might be suitable for Germany, but it was not useful in the desert. I noticed that the water flowed over the stones of the slightly hilly landscape without penetrating the surface. It evaporated too quickly and did not seep deeply enough into the ground to reach the roots of the plants. The seeds we sowed could not germinate or grow due to lack of water. Helmy and I spent days and nights trying to get these Waldhauser machines to operate efficiently, as they were also always breaking down. Helmy was positive that they must work somehow; after all, farmers in Germany were also able to sleep. Eventually we gave up and found a completely different system of irrigation. We terraced the entire ground and dug canals for the water to flow to the fields, like in the Nile delta. Nowadays we have many different irrigation methods in operation on the farm.

Shade-giving trees

I started my next project as soon as I had succeeded in irrigating the fields successfully. I wanted a border of trees encircling the entire plot of land and rows of shade-giving trees lining the roads. Egyptians are not able to experience the wonderful tall trees of European forests, which can inspire people to inner uprightness. They rarely meet this form-giving vertical quality in the vast expanse of the desert and it is all too easy for their soul to disperse in the endlessness of the horizon. In former times the obelisks placed in the temples represented an inner experience of uprightness. This was a further reason why I wanted to plant trees, as well as their ability to give shade and protection from the excessive cosmic influences of the light and heat.

I had planned to plant a small forest in one area of the grounds, but I found that in the meantime the Bedouin had set up camp there. This meant they were in the way, but I was sure it would be easy for these nomadic people to move to another area of the large plot of land. So I went to them and said something like: 'Listen, you will need to move somewhere else with your straw huts.' They suddenly became angry and answered me brusquely: 'No, you need to move away. We are Bedouin, this is our country and you are foreign!' Suddenly they had forgotten that we had been friends up until then, and the mood changed. I had never thought about the fact that this aggressive hardness and rigidity was typical for Bedouin, and stood mystified as to my further actions. I tried to talk to the head of the clan and to my watchman Mohammed — but in vain. He only said with regret: 'Mr

3. The Beginning

Doctor, I love you, but I cannot move as long as the *Gomaa*, the chief, says that we should stay here.' Then I showed them that according to the plans for establishing the grounds water would flow right through their huts. It was no use. They refused to move.

I went to the mayor responsible for this area and asked him for help. He advised me to be careful, as the Bedouin could become very dangerous. They frequently shot people when angered. And he sent me away with a shrug of his shoulders. Then I met up with one of my relatives, and discussed the problem with him. He answered me heatedly: 'Look, I will go to these people and shoot them. You are trying so hard, and if they want to stop you, I will do them in.'

Next I went to the president and explained my problem. He sent me someone to speak to the Bedouin and then reported back to me: 'It's hopeless! We've always got the same problem with the Bedouin. Their stubbornness and rigidity make it impossible to deal with them! We could drive them away, but that will end your peace as they will come back again at night. But I can give you a bit of advice: the Prophet also had problems with the Bedouin, and he solved the problem by marrying one of their women. That led to them being on his side.' Great advice!

But what should I do? Very carefully I tried to get to know the Bedouin better. I noticed that they were actually a pitiful people. Although the men had weapons, they obviously only used them to shoot up into the air occasionally. After some contemplation I hatched a plan which probably seems disconcerting at first to the European mode of thinking: But I was determined to impress the Bedouin with a little scene, without wanting to harm them in any way. With the president's help I received an appointment with the head of the police of the area and asked him for forty to fifty soldiers. He warned me: 'I can send them to you, but I can't be responsible for the consequences which this action may have for you.' One day a truck full of soldiers drove up to the farm and the uniformed men marched in with loud steps. All the Bedouin were shocked. I had built a circular flower bed where the Mahad, the Round House, now stands. My administrator sat there with instructions to give the Bedouin money once they declared themselves willing to move. I sent each one singly to the administrator, and after they had signed for the money their belongings were transported to a new place by a tractor I had organized in advance. There was a great commotion, clamouring and shouting! But despite their anger they did not dare resist because the soldiers were present. I had managed to impress them! They are today still living in the place they moved to.

In the end it was not the soldiers who made them accept me as a life-long, respected authority. After the move they continued their terrible cursing, and they threatened to kill me once the soldiers had left. Even though they had received money for moving they still felt as if they had been driven away. I tried to stay calm and with mixed feelings decided to remain alone on the farm the following night to prove I only had their best wishes at heart. If they wanted to do something to me, they could, but it would be foolish. The courage I showed by staying the night alone in the desert must have convinced them.

At last I was able to plant the forest and the rows of trees. We got 120,000 seedlings: casuarinas, eucalyptus and Persian lilac. It was a struggle transporting them all by car to the farm because of the difficult dirt roads, but with the help of the Bedouin and thanks to the irrigation system in place the plantings were a success. Unfortunately, as soon as the small trees and had started growing, the goats came and ate up all their leaves and shoots. I had to go to every owner individually and say: 'Mohammed, what has your goat done?' and have men guarding the trees against the goats day and night. This sharpened my sense for organization and I started including the Bedouin in the venture and giving them work. Nowadays, they mainly work as shepherds and guardsmen of the land. I was incredibly glad to see the trees, which were very undemanding and did not need anything apart from water, were actually growing!

People of the desert

During the course of time great civilizations have developed along the mighty rivers like the Euphrates or the Nile. The water served to temper the souls of the people and opened them up towards influences from other cultures. Water also led to green gardens and made farming possible, which gave the basis for culture and religion to transform the people living there. But what kind of people went into the desert and how were they shaped by this extreme territory? — The Bedouin as people of the desert are continually subjected to the overpowering influences of light and heat. They have to be able to shut themselves off from their outer surroundings; otherwise they would feel torn apart internally. Desert people have to withdraw into themselves to avoid becoming parched from all the dryness. The liberating, calming, releasing quality of water is lacking in their experience of life. Surrounded by sand and stones, they cannot profit from the mediating green of living plants. On top of that are the harsh contrasts between the deep black of the night and dazzling bright daylight, of cold and heat, which shapes their character. In this struggle for survival, the nomadic people, who have no need for anything, must see all cultural life as superfluous and weakening. The Bedouin do not have art in the western sense of the word. The women look after the home and animals, and they have little time for culture. They spin the hair of goats and camels and weave useful things like blankets for the tents and clothes in colours we think of as garishly bright. Water is too precious for washing and cleaning and is substituted by sand. In general, women are experienced as a burden, as an additional mouth to feed under these harsh circumstances. In earlier times most newborn girls were buried alive immediately after birth. Even today many Bedouin still regard women as insignificant, even though the Prophet in Islam spread a different message.

Nowadays, in my opinion, all the energy the Bedouin needed for survival in former times has been transformed into an extreme shrewdness. In contrast to the settled Arabian population, who have built up their own language, art and moral code, some of these desert people earn a lot of money through smuggling — after all, they have excellent knowledge of the desert. They find it difficult to accept laws which they feel are forced upon them from the outside. With the money they earn they are able to expand their business, and instead of camels they now have four-wheel-drive jeeps. Their settlements in the desert appear cultureless, the houses disproportioned, dark and dirty. They sit on the floor together with their animals. They do not use furniture to eat off or sleep on. Buildings of rich Bedouin can be recognized by their huge entrances, which rise up disproportionately out of the desert surrounding. They usually continue their simple life-style in their palaces, which are often built by Europeans or Americans. The Bedouin only really feel comfortable in their 'huts of hair.'

They meet people like the fellahin (farmers) at the edge of civilization, who live in solid houses and have gardens and fields. The Bedouin, a people given purely to nature, reject this life-style as soft and not worthy of humans and are not interested in assimilating themselves. They realize that people who have settled and established a homestead desire nothing more than peace. So they keep this 'peace' for a price. 'I am your watchman!' Mohammed had said! How naive I had been to be so grateful for his 'kindness'! The Bedouin's reticence reaches very deep. Even their children in our school can hardly stand experiences of culture and beauty which open and expand their souls. We hope that their hard, proud disposition will be touched by sufficient love and attention, even if this task takes generations to achieve.

Energy supply

My main concern during the first years of establishing the venture in the desert was ensuring a constant energy supply for the many water pumps and the electricity production. If a diesel engine failed or a filter in the generator got blocked the farm was plunged into darkness, and, within a short time the ground would become a desert again. It also meant we had to wait until new diesel oil was delivered or spare parts could be brought for the engine. Because of this it was very difficult working solely with generators. We also had some major setbacks with them. We were hit particularly hard by a sandstorm which was so severe we could not see our hands in front of our eyes. It covered everything with sand and the entire running of the farm was brought to a standstill for several days. When something like that happens a drop of water becomes as precious as an entire kingdom!

What I really wanted to have was a continuous energy supply. I looked into different methods and had a few good ideas, including solar energy and wind energy. I ran tests on their implementation, but unfortunately had to give up the idea as useless for this country, it was either too expensive, or we did not have the necessary requirements, as for the wind turbines. So we had to fall back on a state electricity supply with overland electricity lines and transformer stations, and to use diesel as an emergency unit. The nearest place we could connect to was nine kilometres away! All the tasks involved, particularly building and erecting the masts on foreign estates, required tenacious dealings and a lot of work and money. We also fed the people who then laid the electricity lines, mainly thanks to Gudrun.

Gudrun visited me frequently at the farm and saw how everything was progressing. In the first years of establishing the farm she lived in Cairo with the children, who still went to a German school. She survived everything courageously, gave me support and took over the responsibility of the family and school. She had close contact to my family, who welcomed her warmly and loved her. Once our house on the farm was ready to live in, she moved out to the farm and managed the entire household in the first years. When people saw how a European woman could withstand the desert they were encouraged by her example. Thanks to her great energy and strength she was an incentive for many others.

We slaughtered a sheep once a week for the electric line fitters following Gudrun's instructions, and made the meat into a meal. After two and a half years of building the electric line we were able to press a button for light for the first time. We celebrated this occasion with a huge banquet, slaughtered a cow and invited everybody involved in the work with their families. We were very happy with the electricity, even though in those first years there was an eighty percent power failure rate and we needed to use the diesel generators again. Nowadays it is the other way round, and the generators are only used for the cooling and drying machines of the different companies.

The Round House

While drawing up the first plans I already set aside the area for our first living quarters, the Round House, which is used as a guesthouse today. I worked mainly at night on its conception, as my daily duties were so excessive they left no time for additional planning work. I thought its outer curves would be able to shield the house from the openness of the desert, enabling a protective centre for cultivating plant life and also cultural and social life. I wanted to plant a garden with scented jasmine, hibiscus and roses in this miniature oasis in the midst of a barren surrounding. But what materials should I use to build the house? I looked at the houses in my surroundings and discovered that they were made out of clay, which is first stamped on and then built up by hand. So I employed some builders who gathered the clay with their donkeys; we already had the necessary water on the farm.

Next I showed the men my plan, which I had carefully drawn for them on a piece of paper, and explained the quite simple construction. The three ground circles were quickly measured, and they started working. But already by the second day the paper I had given them was completely tattered and the plans were unrecognizable. They had kept looking at the paper with their wet, dirty hands to see what to do. They could not continue like that. So I drew a new plan, which I fas-

The Round House, the first building at Sekem in 1978.

tened to a board and forbade them to touch it with their fingers. They should only look at it. But after two days the board was gone, and the paper was dangling from a bush in the sun. Finally I took a cloth and drew the plans on it with ink. The builders were obviously relieved! Now they could put the plan in their pocket without worrying about it, and even use it to wipe the sweat from their brow. I had managed to give them a truly practical tool!

The builders thought it was a strange concept to build a bathroom in a living house. It was not part of their culture. With great difficulty I acquired red bricks to attach the tiles to, and I had to oversee and correct every step of the way. I hired a joiner to make the windows and doors according to my drawings. Electricity, telephone, fresh water and sewage pipes were all put in place — what a lot of work in the eyes of the workers! I could tell they were wondering what the point of all this was! Once the house was finally finished, I took all the workers to Cairo by bus and showed them what the living quarters looked like there. Then I showed them around the zoo as a reward for their commitment.

3. The Beginning

Economic beginnings

The biggest question was how to finance the whole venture. Even if we had managed to finance everything up to this point, the grounds were basically still a desert. Where could we obtain a new source of income for houses, plants and animals? I realized we needed something in addition to all that had been created for the farm, i.e. businesses, in which people worked and earned money to finance the cultural institutions I had planned for the distant future. I tried to find out how I could use my pharmacological knowledge to produce things for the people of Egypt and for the export market. It was time to get off my tractor, don a suit and tie and drive into Cairo to talk to people.

I went to visit Ahmed Shauky, my father's tax consultant, and asked him to take over this task for Sekem. I explained my vision in the desert to this elderly, distinguished man. He turned out to be delighted and very interested. During our talk his son, who had been followed our discussion attentively, said: 'I have heard that an American business is looking for an extract of the plant *Ammi majus* from Egypt. Maybe you could do that!' I immediately ordered a report from the company and invited the Americans to a meeting in Egypt. Up until then I had never heard about this plant nor did I know how to get the extract. The company only wanted the crystallized active ingredients: ammoidin, which is present in the seeds and is a medicinal herb for healing skin pigmentation disturbances.

I needed to start learning again. I remained in the library for hours until I had found out about all the necessary information. *Ammi majus*, known in English as Bishop's Flower or Laceflower, is a wild medicinal plant which grows both in the desert and in the delta. It is an umbellifera, about as high as fennel or aniseed and grows as a weed in lucerne (alfalfa) fields. I observed the *Ammi majus* seeds exactly so that I could explain how they were different from other seeds to the those doing the sieving.

I planned the buildings for the extraction plant for many nights and as part of this process became acquainted with Hassan Fathy, among others, who was awarded the first Alternative Nobel prize and is known for his traditional clay building architecture. I deliberated what the machines should look like and calculated the cost of the project – and I realized it could become a lucrative business. So I started building the workshop, bought stainless steel and constructed machines for the venture. After we finalized the contract with the American Elder company in Ohio we had camels and trucks with sackload of *Ammi majus* seeds coming to the farm for years.

I wanted to enter into a partnership with a bank for this huge project which I could not finance myself. I chose an Islamic bank recommended by a friend as a co-investor, as I assumed it worked according to Islamic principles. Allah says in Islam that the earth and the ground are only given to us to care for. He alone owns the ground. It is the same with capital, with money, we can only manage it for the good of the people, but should not call it our own. He says that whoever enters into trade works together with Allah and following his principles should give the proceeds to the poor and needy by giving up his own possessions. In the light of this Islamic esotericism I perceive modern joint-stock companies as dysfunctional, as they act as if God's legacy were their own. The interest and the resulting riches they receive are not their own achievement, because even intelligence and individual abilities are the gifts of Allah, even if modern humans think they are solely due to their own efforts.

These Islamic ideas appealed to me, particularly the idea that money is not a commodity which can be bought and sold again with interest. Thus I was happy to have found an Islamic bank where I would be able to work together in a like-minded partnership — as I hoped then. It turned out the practices of this so-called Islamic bank were the same as any other money institution.

The Sekem Company was established as an investment company right at the start of Sekem. Because I needed at least three people to start a company according to Egyptian law, I included Helmy and Mona in the business, even though they were still underage. The bank wanted to inspect everything and I had to open my books for them. The negotiations were tough, and only succeeded once the director of the bank had started sympathizing with the idea of Sekem. We agreed on a forty percent share of the bank. Because Sekem was seen as a foreign company the state had the right to say something in the matter. The company itself was protected, but the state had to give its permission to the bank. The bank agreed to this, got a provisional authorization for the project with *Ammi majus* and signalled we could go ahead with the project. I ordered the first extraction machines from Denmark which the bank paid for. After some time the state investment authority requested accounts from Sekem. It did not want the book value, but the estimated value. So an estimation committee worked on the

3. The Beginning

farm for several days, re-examined everything and found that the estimated value lay way above the book value. This meant the bank had to pay a higher amount for involvement in the project. But the bank was reluctant to accept this, beginning to doubt everything and trying to get out of the contract. They demanded back the 150,000 pounds they had already paid out for the Danish machines, but I needed this money for developing the farm, and I did not have that amount of money spare anyway. This started off a protracted struggle. An arbitrator was employed and it took months for our two lawyers to decide on a third party to mediate.

During this process there was a small event which happened by the side, but which sheds light on the way the negotiations were held: one day my lawyer came to me and said: 'Listen, if you give the bank's lawyer 10,000 pounds then he will accept the estimated value.' — 'My friend,' I said to him, 'You know me. I will not pay bribes. It does not correspond to Islam!'

Once you have a dispute with one bank, all the other banks and the central bank know about it. This meant I was always rejected when I attempted to find a new investment partner for my project. The banks always told me to settle my disagreement with the Islamic bank before further negotiations with them would be possible.

Then one day a relative visited me and introduced me to an Egyptian who had just come from Saudi Arabia and had a lot of money. He thought he would be the ideal partner for me. The man was called 'Mohammed,' was inspired very quickly and invested 100,000 pounds. But after only two months he came back to me with the excuse that his wife wanted to go back to Saudi Arabia and he needed his money back immediately. I had already spent his 100,000 pounds on an important machine and could not give it back to him immediately. The debts and conflicts grew! I had met another 'friend' during my search for suitable partners, but they were all people who did not understand my vision and only wanted to make a quick buck.

It now looked like the *Ammi majus* project might fail, and with it the farm's survival was threatened. I decided to put all my eggs in one basket and went to visit the director of the Egyptian National Bank. I explained everything to him, and ended with the words: 'If you do not help me and lend me money against the security of the land and houses on it, the project will die!' The director of the bank could immediately see that his money was covered by the contract with the American company. There was hardly any risk involved for him, and he decided to finance the project. 'Don't worry about anything else, it's all settled,' he ended. At long last we could start the contract with the American company.

The dispute with the Islamic bank was only resolved years later. We had suffered great setbacks by its pulling out, as it stopped us entering a new partnership, and instead we had to get a loan. In the end we paid them back three times the amount just to finally have peace. We had to give a piece of land to Mohammed from Saudi Arabia, who was demanding his money back with threats. This seemed like a great loss to me in those days. By now we have bought back most of the land apart from a small section and the dispute has been resolved.

With the money from the National Bank I started building a laboratory and the processing rooms for extracting the active ingredient ammoidin. I received the conditions from the American company on how to deliver the substance and my former knowledge in technical chemistry turned out to be very useful. By and large I did all the experiments necessary for the production process myself.

We needed a steam generator for the extraction which was very expensive. Then I discovered an old German wood-powered steam locomotive at a scrap dealers. I had it disassembled and bought the steam-boiler, which still stands at the back of the farm as a kind of museum piece today. The extraction building also needed a chimney: thirty metres high and forty centimetres circumference, for which single pipes, each four metres long, were placed on top of each other. I planned this undertaking carefully: we built wooden scaffolding so that the workers could pull the pipes up with ropes and place them on top of each other. But after only ten metres the scaffolding started swaying and everyone ran away! Apart from Helmy, who bravely continued helping me with the building. He always stood right at the top and had the pipes handed to him, and encouraged the others to continue by his example. I supervised the building process and kept the overview. When people were left to work by themselves, which was also necessary, accidents always happened. For example we bought a tank for the diesel oil needed to power the steam engine. For cost reasons we purchased an old tank, had it cleaned inside and out and painted new. The man who sold us the tank wanted the work to be carried out on-site. A young man entered the tank to clean it from the inside with petrol, and then lit up a cigarette in his break — with drastic results.

Such tragedies happened repeatedly when I was not present.

3. The Beginning

Another time a neighbour wanted to divert electricity for his radio from the 11,000 volt power line leading to our farm.

Once the sacks of *Ammi majus* seeds had been delivered to the farm, the production commenced as follows: all the seeds were separated carefully from other seeds by re-sieving them. Then they were ground into a powder and put into percolators, stainless steel boilers. Water and alcohol dripped in from above, the active ingredients were extracted in this manner and came out as a brown soup below. This alcohol-water-mixture was evaporated with the help of the steam engine. Several cleaning processes followed so that the crystals were completely pure.

For years we worked well together with the Americans, until one day I received a phone call from the Elder company in Ohio asking me to come and visit them. Once in America I was offered the company to buy. I was told the director had died and his children were not interested in continuing his business. They were asking for a reasonable price, but unfortunately I did not have the money, particularly as they had failed to pay regularly towards the end. So our mutual business ended, which, despite initial difficulties had helped me further in my phase of establishing the farm, and I now had to find a new line of business.

Changes in the family

Once Helmy finished school he remained on the farm for a year to help me look after the animals and with the irrigation system, while I concentrated on establishing the industry and marketing. Then he went to the Waldorf teacher's training course in Witten-Annen, Germany, for a year. He had become acquainted with a girl called Konstanze in the German school in Cairo. After she returned to Germany with her parents Helmy visited her several times from Witten and a year later they got married. On our return journey from America Gudrun and I went to the wedding, which Konstanze's family had arranged with loving care. Singing, dancing and recitals lasted till late at night. I can still remember the beautiful dress that Konstanze wore. Once she came up to me during the festivities and whispered confidentially: 'Do you see the woman over there? That's my grandmother. Can I introduce you to her?'

After I had greeted the old lady, Konstanze returned to me and whispered again: 'Do you know what she said before? She asked me how old Helmy is, and on finding out he is only twenty, she asked if she could tell her friends he was already twenty-one. And I said she could. I thought I should tell you so you won't be surprised to hear his age has changed.' I hugged her and laughed. What a wonderful, cheerful lady, and she has remained so until this day. Both of them moved to Sekem after their wedding, where we had further celebration with all the co-workers.

Helmy has been involved with all the problems on the farm and acted as a 'chemist' for producing ammoidin. He has become a great support. Our daughter Mona worked on the Sekem farm for a year after her school leaving exams and then went to a curative education centre for practical work experience at Lake Constance in Germany. After she returned she helped us establish an adult education program on the farm. She is a very talented teacher and adults enjoy working with her. She also continued to work on the farm as well as her teaching duties. Then she went to Hamburg to study eurythmy. Now she lives near Freiburg, Germany, with her husband and five children. Her husband manages the German Association for Cultural Development in Egypt.

Increasing duties

After the business with ammoidin ended I started looking for a new venture in Egypt. I had the idea of making aromatic medicinal herb teas to remedy common ailments occurring in Egypt. I designed small ten gram sacks, which were initially filled with the herbs by hand using spoons, and called these teas 'Sekem Herbs.' In those days we started planting medicinal herbs like chamomile, peppermint and mullein on the farm. I submitted the recipes of the teas to the Health Board to have them registered.

Now I discovered what it was like to market internally in Egypt. I visited doctors and pharmacists to promote the new products. Once I noticed that the marketplace demand was increasing, Helmy and a friend invented a filling machine with a hydraulic pedal: a small plate rotated slowly, from which the herbs trickled into the sacks. We employed women for these tasks. Once the medicinal herb teas were a success on the market, we expanded the range to further herb teas. At the same time we sold our agricultural produce like milk, cheese, bread and vegetables to the German embassy and to the German

3. The Beginning

schools in Cairo. This was Konstanze's task. Through these activities we gathered experience with the Egyptian market: What can we sell? Will we receive the money? Will retailers re-order the products? Extensive planning with marketing and economical viability studies started each new project.

During the first year we did not have a telephone on the farm and our electricity supply was unreliable. But because I needed some form of communication to be able to manage the different registrations, the sales and hundreds of different inquiries I set up an administration office in Cairo, for which I had already bought a plot of land. Setting up this administration office turned out to be even more difficult than anything I had ever experienced in the desert. I had to accept that the people employed in administration worked terribly incorrectly and unreliably compared to European standards of punctuality and organization, so that initially I was forced to supervise them continuously. This was not easy beside the many other tasks I had to perform and I desperately needed people to take on some of my workload. On top of this, supervising adults, of whom I expected more independence, went against my moral sensibilities. And yet, over time, it turned out that this was the solution for this country: people were very grateful to be inspected; they felt it enabled them to follow a path they could not keep to themselves. They never reacted with rejection to corrections; indeed the opposite was true, they experienced them as assistance towards achieving success. They felt noticed and accepted when supervised. Few people in Egypt possess abilities like planning and self-control, most of them live into the day without much thought. I tried to set an example for the people I worked with so they could acquire these qualities through imitation.

As I only gradually discovered the importance of constant supervision I was often ripped off because I was so trusting. For example, to establish the Egyptian market I needed experienced and reliable people working outside the office and as salesmen. Much depended on their trustworthiness: whether they really brought the goods to the pharmacists or quietly let them disappear at home, or sold them to a pharmacist who then discontinued sales and I never received any money. My salesmen had to deal with money a lot, but in this country whoever has the goods and the money think they can do what they want with them. They could for example say they never received any money and then keep it for themselves. This has all really happened! So I had to establish a system of book-keeping in a country where businesses do not keep their paperwork in such an organized way as I was used to. I noticed that the pharmacists hardly ever kept to arrangements. They were all miles ahead of me in their shrewdness. I was trying to make plans in a country which has no time or feeling for order or for keeping to arrangements.

We also needed to build an office for the administration in Cairo, for which I hired a building firm. The building team demanded a certain sum in advance. Then they stopped working and disappeared — with the money! I was under the impression that advance payment was usual in Egypt, and did not want to take the risk again. Then a friend drew my attention to a building constructor whom he said was very honest. I met up with the building constructor and complained about the people who were crooks in Egypt. He answered: 'Yes, I have also come across them. But you do not need to pay me a penny in advance, only once the house is finished and you are holding the key in your hand.' And he quoted an acceptable price for the whole building. I was relieved. At last I had found somebody who was reliable, and I handed him the plans which I had again drawn up during strenuous nights.

I was relieved to see twenty people digging the foundation when I drove past the building site the next day. The builders were still busily working the following days and I revised my opinion about the dishonesty of Egyptian building contractors and was placated. Every day they continued working diligently, leaving the building site tidy at the end of the day. Everything seemed to be working well!

One morning a week later I saw the twenty workers sitting around doing nothing. I went up to the foreman and asked: 'What's happened? Why aren't you working?' — 'Our building constructor is in Ismailia and will be coming back presently,' was the answer. But two days passed, during which time nobody worked and the building contractor did not come back. Gradually I got impatient. Then the foreman said to me: 'He's probably been held up in Ismailia. If you can get us iron, sand and cement then we can continue without him.' So I instructed my driver to buy the necessary materials with the foreman and gave the driver sufficient funds to pay for everything the workers needed. Then I went home and forgot all about the issue as I was occupied with other things. Shortly after midnight my driver appeared on my doorstep, wailing loudly. When I asked him with surprise what had happened, he told me how the foreman had taken him to a shop he knew. He asked for the money to pay the shopkeeper for the goods.

3. The Beginning

Then he entered the shop — but never came out again. The shop had had 'two doors'! I never saw the building constructor, the foreman, or the money, ever again. Luckily Sekem now has its own trustworthy building team.

Because of such unreliability people meet each other with constant mistrust when dealing with business matters. One of the important tasks of Sekem is to consciously trust people with their work so that they can reveal their honesty. This is awarded with 'supervision,' which the people here experience differently from Europeans, as already mentioned, as being seen and taken into account. This task will take decades for lasting results to be achieved, but a new morality in dealing with each other can be developed. It has to start with little and requires daily attention — but it is a worthy task!

Opposite: Ibrahim Abouleish in 2003.

4. Prevailing

Resistance

While I was still studying in Graz I had been invited by the Egyptian embassy in Vienna to talk to young Egyptians about my life and to philosophize with them several times a year. A circle of people who got on very well together and supported each other developed out of these meetings. Many of these young people returned to Egypt after their studies, others were only in Austria for a short period of practical training in the first place. As soon as these former students heard I had returned to Egypt they contacted me and we began meeting again regularly once a week in Cairo. The type of meetings remained the same as in Vienna. I told them about my vision, reported about Sekem, explained Rudolf Steiner's view of the threefold social order and lectured on stimulating themes from anthroposophy, which we discussed in detail. This circle was extremely helpful and important for me during my setting up phase, as the participants helped me to make contacts and become known in the country. In addition, farmers, administration specialists and enthusiastic young artists came to the farm and helped me establish it. Among them was a young music student, Osama Fathy. He brought the first sounds of music into the desert with his Egyptian string instrument. He also organized the first grand piano for Sekem. It stood in the Round House for many years and around it we sang and played music.

I noticed I had the gift of inspiring people. They wanted to work well because they were infected by my idea, or for my sake. But outer circumstances often conspired against well meaning thoughts. It took hours alone just to bring workers to the farm from Cairo. I purchased a bus for these transports, but the return journey took forever because of the bad condition of the dirt roads. The asphalt military road left over from the First World War was in an even worse state, it was full of pot holes. There were many accidents on it and our car was permanently broken.

I often felt very lonely during this period of establishing the initiative. Many enthusiastic people who I would have liked to keep permanently and had become attached to left me soon after starting. They could not stand it here: they felt the farm was at the back of beyond. Some people did not have sufficient life energy to implement their ideas, and quickly gave up because of this.

Most members of my large family were simply opposed to the idea of my initiative. They could not understand my plans. If a successful pharmacologist desperately wanted to become a farmer, then there was plenty of fertile land in the delta. Why did I choose the desert? They could not fathom this mystery, and from their point of view their doubt was justified. Because of this they kept their distance and did not support us, as well as giving me the feeling that the project would never work. The question remained with me throughout the first years: was it the right decision considering all the difficulties? But by the next morning my doubts had dispersed and hope returned. 'Everything will be right. Just wait!' I told myself.

Administration in Egypt was extremely complicated and tedious in those days when I was trying to start the initiative, as indeed it still is now. One time I was supposed to explain biodynamic agriculture and composting to the Egyptian agricultural ministry. When they read my explanations, they decided to ban the project on the spot. What had I done wrong? After I questioned them persistently they revealed that according to my description bacteria multiply in the compost, and they were worried that we would infest the whole country. They could not permit something so irresponsible. It took weeks to correct this disastrous mis , and to even get the professors and administrators back together to negotiate with them. Then I was told I did not know anything about agriculture as I was not a farmer. So I had to argue, bring literature, explain the process of composting and exactly what how it happens. I started studying throughout the night so I was able to offer the answers. By the morning I had all the answers ready and gradually I was able to persuade more and more people to trust me on the topic of composting. But I had to work singly on each person! I learnt a lot during this process. But the decision was still postponed. Meanwhile I continued working on my project in the desert, until one day the police arrived, stating: 'You are not allowed to continue working!' They declared it was not clear yet whether permission would be granted at all.

For nearly twelve months I had to struggle with huge difficulties, until unexpectedly it all suddenly changed. The ministry let me know they were going to send inspectors out to the farm to see how the earth evolved after treating it with compost. A scientist came and took a sample of earth to analyse the ground. This process was repeated regularly over a period of ten years. And in the end that was the best thing that could have happened to us, as the ministry was able to see the ground improving step by step due to our methods. I won many friends from the ministry and never tired of talking to them about my ideas and vision for the country.

Despite all the resistance my vision of an oasis in the desert, from which I could draw water for plants, animals and humans, slowly began to take shape. I became acquainted with the professor for agriculture, Chairy El Gamasy. He found cows from his native village for Sekem. He also had a field of roses planted for me, because I needed beauty around me in this arid wasteland. We started off with fifty

Above and overleaf: Dovecots.

to sixty Egyptian buffalo. They required so much food that to start with the whole farm was used for cultivating forage plants, which only grew sparsely on the stony ground. I remember how surprised I was by how much food these animals needed. I felt their care was very important, and I did not want to give this task to just anybody. It depressed me that I was unable to find anybody who was capable of dealing with them properly. Finally I assigned the task of looking after and feeding the cows to Isa, the young man who had helped building the Round House. I myself needed to spend more time with the composting process. The Bedouin were not able to look after cows as they did not keep animals in stables, and it was not easy finding fellahin to do it. The cows needed a stable to give them shade and protection. But what should stables look like in Egypt? European style stables were too costly for our circumstances, and we had to make do with provisional solutions for some time. How I longed for someone who could look after the animals in a professional way.

It required a huge amount of effort to set up the compost, which is the basis of biodynamic agriculture, and the people around me could not understand the sense or point of it. Why bother with it all? One

4. Prevailing

could just put the animal manure directly onto the earth, so why have the added work of layering and re-layering the compost? The ideas behind it were completely foreign to them. I had taken biodynamic compost preparations from Georg Merckens with which to treat the compost. I felt preparing the compost heaps and taking the manure and horn silica out to the fields were important tasks, and made sure they were performed with care. Sheep and pigeons joined the cows we owned. The tall, white dovecots belong to Egyptian scenery. They were the first buildings in the area, apart from the stables.

Gradually all the tasks were working well together. The trees we had planted were three years old and had grown to a good height which the greedy goats could not reach any more. But one morning, when I drove on to the farm from Cairo as usual, I could not believe the sight I saw: bulldozers were pulling down thousands of trees. I was met by soldiers with machine guns and suspicious eyes. I found out that a general had ordered our grounds to be made into a military area, even though it was only through our efforts that there was even a water supply on our land. Without further negotiations they wanted me to leave. I felt this was like a declaration of war! My violent choleric temper emerged, and I was able to stop further destruction for the moment by my verbal protests and decisive stand. But I had to go to Cairo to start diplomatic and political negotiations to obtain a more lasting solution.

I already had to spend days in Cairo setting up the administration office. Now it was necessary to abandon my direct work in the desert for a time to fight for the continuation of my project. The Egyptian President Sadat was a good friend I had got to know during my adolescence, so I went to visit him. In the government palace I also met the minister Shabaan, the head of office of the then Deputy President Mubarak. I explained everything that had happened, and he promised to help me. I was so angry and upset that I made everyone's life hell and repeatedly visited or phoned the minister to hurry up the resolution. It still took weeks before the whole military machinery moved off the farm. The concept of compensation does not exist in Egypt, the best that can be expected after a mistake has been committed is an apology. The responsible general excused his behaviour, executed solely on his own authority, and I accepted his apology. Later he was transferred to another area. His successor, General Ali Siku, immediately became my friend. We visited each other and became acquainted with each other. Together we established a cooperative with single plots of land for officers on three thousand hectares of desert. I had

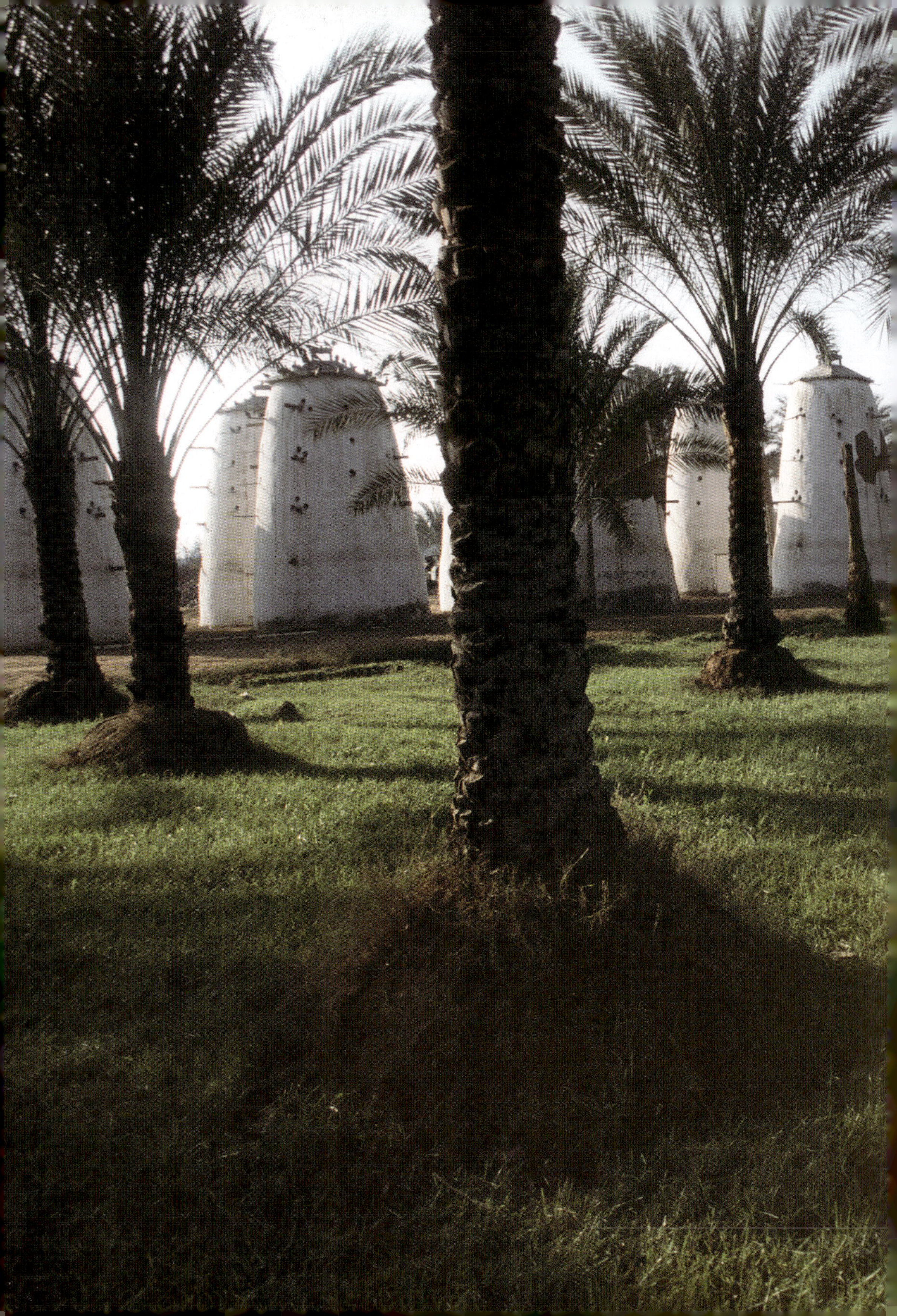

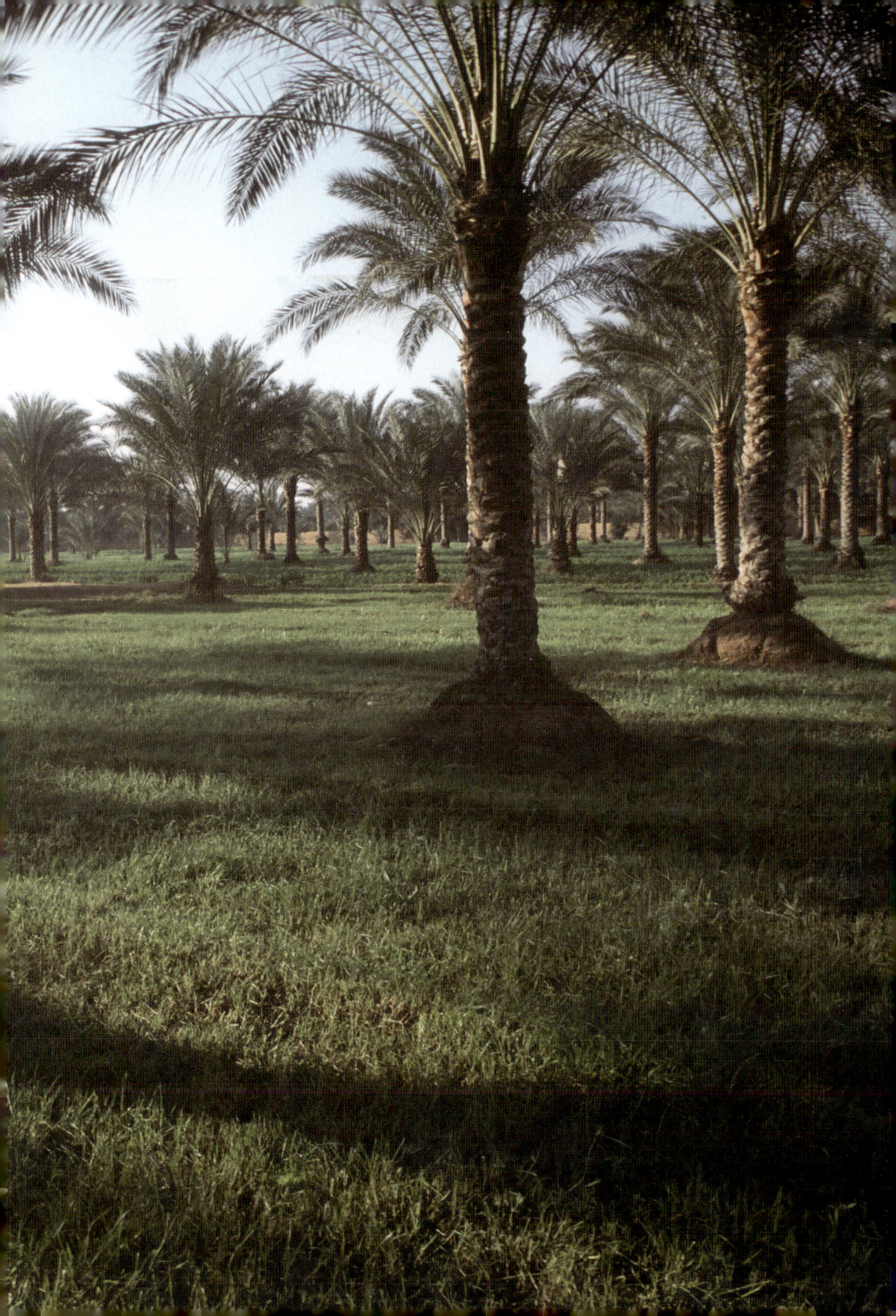

discovered that this had been the original plan of the transferred general, and he had wanted rid of me to implement his idea on my grounds. Now I followed up this idea and discussed it with the new general Ali Siku. I explained it was not necessary to start this venture on the same grounds I occupied and had established into a fertile area. Eventually we agreed on this point and become good neighbours. I helped him establish the cooperative materially and conceptually. The land surrounding Sekem was divided into small plots of five to ten hectares for each officer. The green cultivated countryside seen around Sekem today belongs to this cooperative.

Despite the opposition, I also experienced moments which gave me courage and spurred me on. Since my adolescence I have been working regularly spiritually and through this gained great spiritual energy. I always had a deep inner desire to keep the praying times and meditate on the verses of the Koran, particularly the ninety-nine names of Allah. After I encountered anthroposophy I started studying it as well as continuing my meditations and prayers. When I read that for some people everyday life constitutes a more or less unconscious 'initiation,' and that suffering, disappointment and failure can be seen as a chance to strengthen courage and inner steadfastness, then I felt that the obstacles I encountered were not sent to destroy me, but to steel my resolve. Such resistance had to be met with greatness of soul and continual energy.

The presence of nature also gave me strength. The bushy dark green leaves of the trees were gradually starting to enliven the desert ground of the farm. I could always find beauty to admire: sunrises and sunsets, sparkling stars in the night sky, or glittering dew drops on the leaves. I observed the number of insects and birds increasing on the farm, attracted by the trees and the treatment of the earth. I felt Allah's creative omnipresence through bird calls and animal sounds, smells, the wind and in the blossoming and flourishing around me. The Koran relates how Adam and Eve lived in paradise before satanic whispers led them to the forbidden tree and they were expelled. But the Koran promised to return the Garden of Eden to believers as a most beautiful reward for their devoutness — the god-fearing will live forever in gardens. 'Gardens, in which rivers flow' is mentioned more than thirty times in the Koran. The greatest source of joy for people living in arid surroundings is green gardens, shady oases and flowers and trees. It also gave me the greatest fulfilment to watch Sekem flourishing.

4. Prevailing

Encounters

For years I longed for co-workers who knew how to farm. I realized that I would not be able to implement my vision of establishing a garden in the desert alone, but needed people with different talents at my side, each with their own abilities acquired through life experience. I often felt terribly amateurish. I longed for a farmer, an animal breeder or an engineer to come to the farm and work expertly without supervision. Setting up the irrigation system would be enough work for one person alone. Looking after the animals, the compost heaps, tilling the land, making the milk products, building up the pharmaceutical business — with the related processes of authorization, administration and marketing — and planning and building the houses: the amount of tasks was becoming endless. But the experts to whom I could have handed over some of my many duties never applied for work on the farm. Day turned to night while I worked without much sleep. Despite this I still felt sufficient energy to continue doing the necessary work in those first years. I was never seriously ill in my entire life, apart from a heart disorder, which occurred, I believe, because of the continuous overwork and the growing pressure from outside. I frequently saw people literally collapsing with exhaustion, while I still felt able to continue. I experienced this strength as a gift, for which I am extremely thankful, but I sometimes wonder where it comes from. Every day I spent some time engaged with anthroposophy. I felt if I managed to work even half an hour on something other than my normal tasks then I was refreshed. Despite the strain, my main interest and care was still directed towards the people working at my side to establish the initiative. I believe this gave me the energy to continue.

I always found new helpers willing to work for a period of time on the farm through direct encounters. After three years, in autumn 1980, Monika Kuschfeld, a teacher, visited me from Germany and offered me her help during her one year sabbatical. She was interested in social questions. She lived on the farm for several weeks, and kept asking me what I actually planned. I told her about my vision: doctors, pharmacists, teachers, farmers, book-keepers and engineers coming together to establish a initiative offering education and work opportunities for the Egyptian people. Monika Kuschfeld could not quite imagine my idea, as nothing of it existed yet in reality. Eventually she left the farm to visit the Valley of the Kings to become acquainted with ancient Egypt. In the columned hall of Karnak she met Elfriede Werner and

Frieda Gögler, who were visiting Egypt with a travel group from south Germany. She went up to them and told them about an Egyptian man who was establishing a farm called Sekem in the desert to the northeast of Cairo, and was desperately looking for people to help him with biodynamic farming, pharmacology and medicine. Despite many difficulties Elfriede Werner managed to organize a group trip from Cairo to visit us as part of her journey. So one midday at the end of January 1981, ten travellers turned up at Sekem. I welcomed the visitors with open arms in front of the Round House and said spontaneously: 'I knew you were coming!' Elfriede and I both experienced this as a very special encounter, and the intensity of the meeting was shared by her husband, Hans Werner, a doctor, and by Frieda Gögler. Elfriede recognized in me the image she had dreamt about the night before. And I also did not feel she was a stranger, but a deeply intimate friend. All three spontaneously decided to help Sekem. Once back in Stuttgart, they told their friends about their experiences. They immediately tried to inspire Angela Hofmann, an experienced livestock owner, to work in Sekem. That same autumn my son Helmy and his wife Konstanze decided to return to Sekem after their honeymoon.

Things happened that year which for different reasons made me think. Before Angela Hofmann decided to come Elfriede arranged for Martin Albrecht to join us, a young farmer from Pforzheim. He was a cheerful, competent young man, who was soon able to work independently on the farm. But after one month he became seriously ill. Hans Werner, who was visiting at the time, cared for him medically. After a few day we had to take him to hospital in Cairo as he had such a headache and temperature. Diagnosis: myeloencephalitis, severe meningitis. We were all extremely worried about him. Gudrun looked after him day and night. After three days he died. His death left a great hole. We were all shocked and followed his repatriation to Germany. Helmy and Konstanze went to his funeral in Pforzheim.

When Angela Hofmann came she immediately began working with the animals with great enthusiasm, set up the cheese dairy with Konstanze and took charge of the bakery. She made sure that forty cows from the Allgäu were shipped to Egypt. Their manure was excellent for the composting. One can hardly imagine how glad I was to have a reliable co-worker, who could be an example for the Egyptian employees on the farm.

During the year 1985 I noticed my verve for life which had kept me going up until then was gradually waning and I was starting to reach

Angela Hoffmann, Eissa, and the first Egyptian cows.

my limits. The accidents on the farm began to increase. One of them was as follows: I desperately needed a warehouse for all the herbs that were transported to Sekem. I heard that there was a leftover English warehouse for sale in Cairo. It was 40 × 60 metres, an enormous 2400 square metres! I became acquainted with the architect Winfried Reindl from Karlsruhe through Roland Schaette. Reindl came to Sekem with a group of enthusiastic young people to help with architectural planning work. I asked him how and whether this warehouse could be erected. He went to inspect it, and returned slightly discouraged, as the building had not been constructed well and needed massive foundations. But we began digging the foundations according to his plans, while the warehouse was dismantled in Cairo. It was extremely difficult taking the building down and transporting it. But as I was busier than usual with office duties at that time I was not able to supervise the construction daily.

One day, after the building had been roofed, I was disturbed by a telephone call in my office: 'Dr Ibrahim, come quickly! The warehouse is in ruins!' When I rushed outside I saw everything scattered over the ground. The whole building had collapsed like a card house

Dr Aboulesh receiving the Alternative Nobel Prize, the Right Livelihood Award in Stockholm December 2003 on behlaf of Sekem.

Dr and Mrs Abouleish with children and grandchildren.

4. Prevailing

Dr Ahmed Fathy Sorour, leader of the Egyptian Parliament, at Sekem.

The Sekem headquarter building in Cairo.

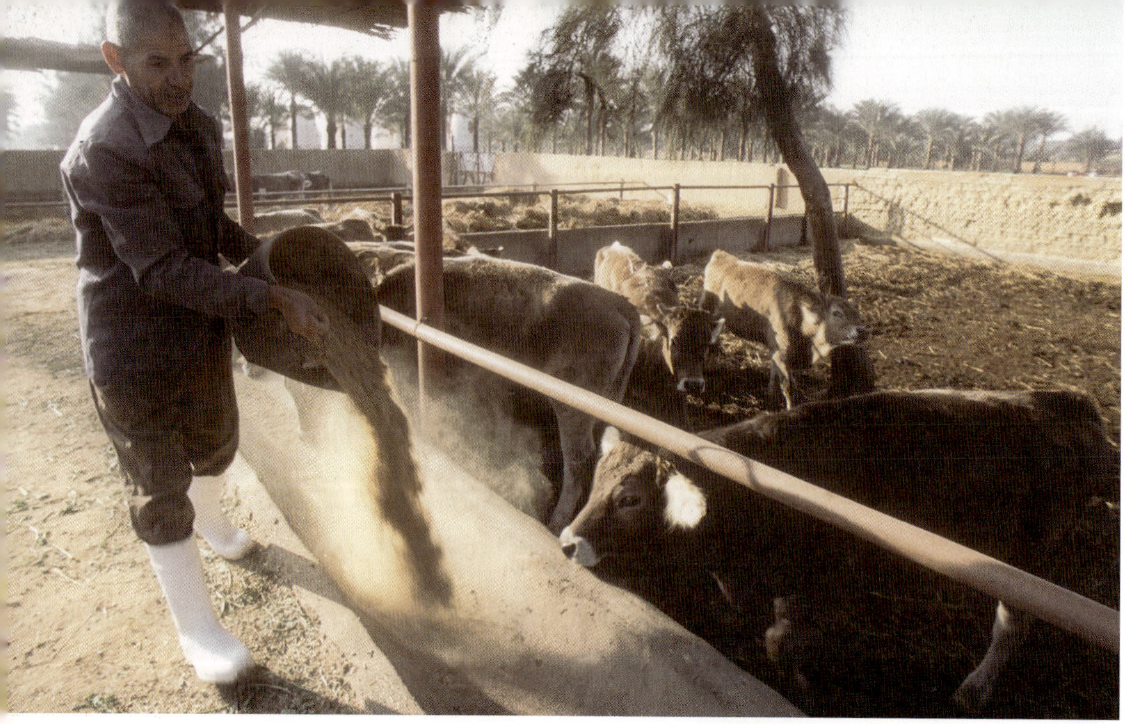

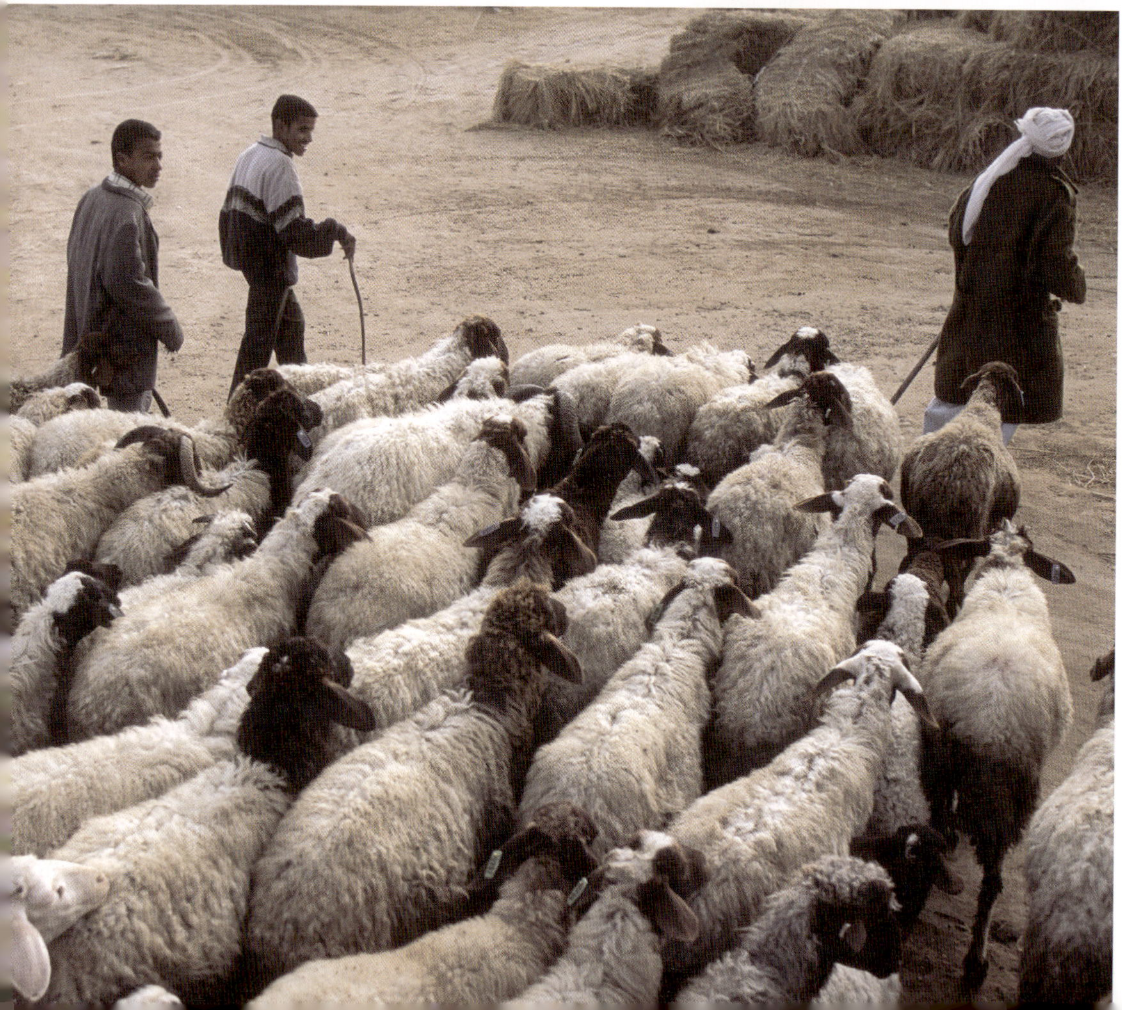

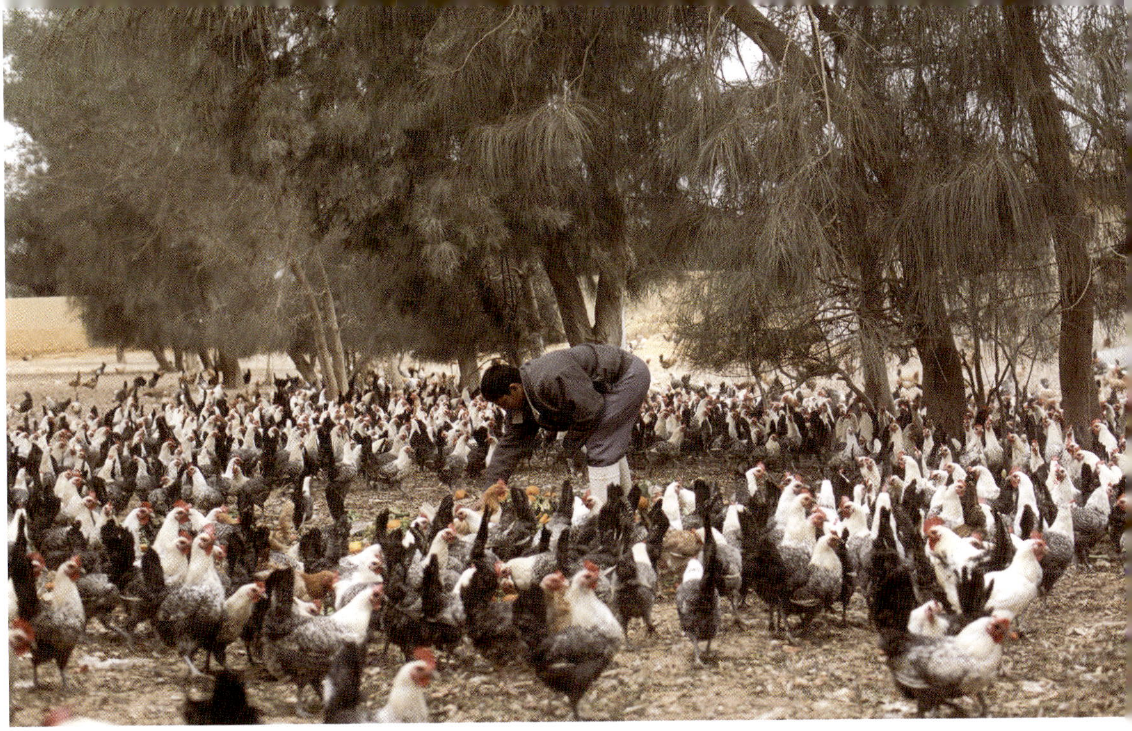

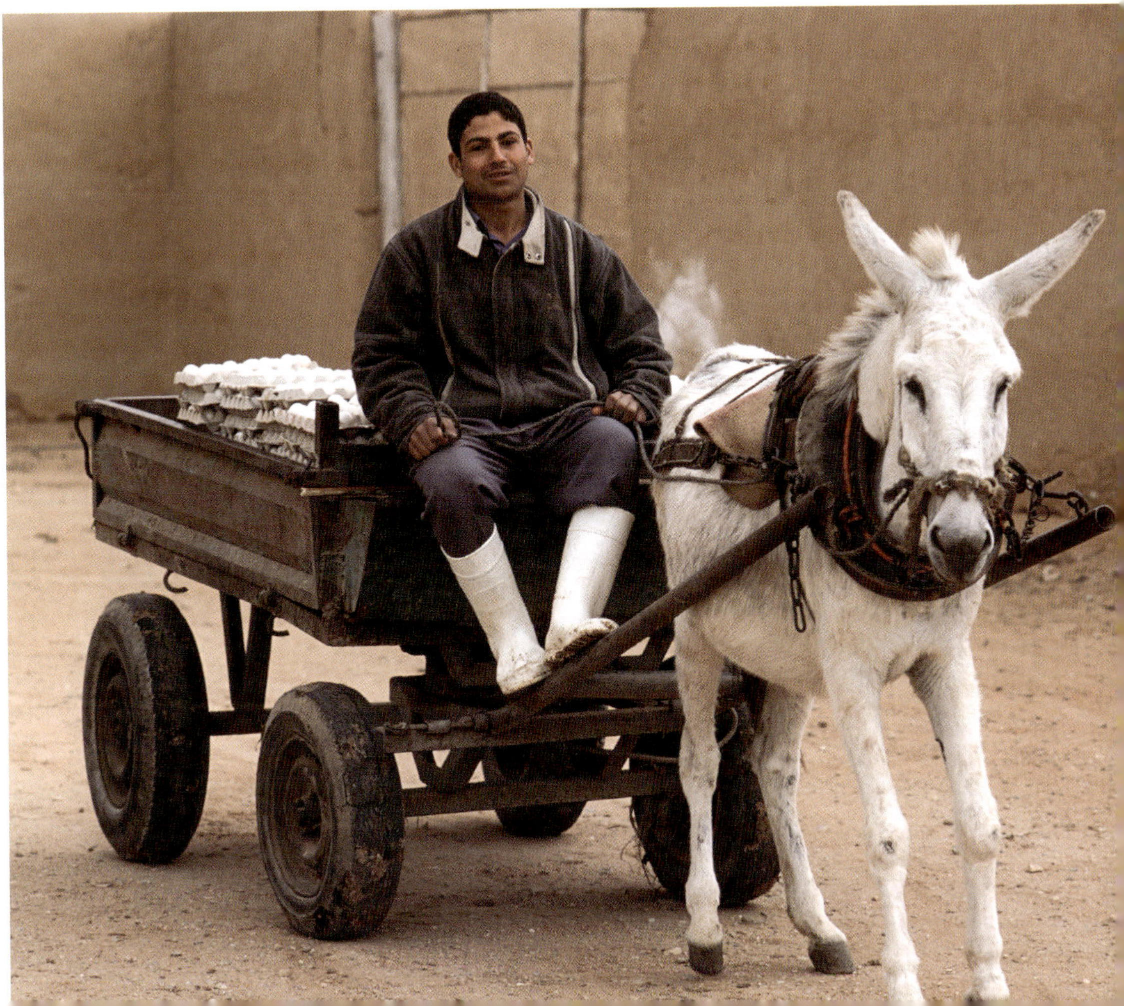

and the people were all shouting in confusion. Thank goodness no one was hurt, as the building had been empty when it collapsed. The eight metre high iron poles were useless now. After some enquiries I found out that for some reason the builders had not tightened the screws at the base of the iron poles sufficiently when erecting the building. In Egypt you cannot make anyone take responsibility when such disasters occur. Because of such occurrences I realized that I wanted to achieve more than I was able to do alone, and that I needed people who could support me reliably.

At that time there was one ray of hope: my daughter-in-law was expecting her first child. Every time we saw each other we tried to find a name for the child — she was hoping for a girl. On February 4, Sarah was born. But in general I felt my vision becoming obscured due to the tremendous demands life placed on me. The obstacles mounted on all sides, causing great pressure on my soul. For months I could not find a project to give me hope and where I felt I could continue. Added to that were many disappointments in the people around me. Nothing turned out as planned on the farm, and I would have needed to be present everywhere at the same time to prevent disaster. I was busy fixing one catastrophe after the other almost twenty-four hours a day. Because of all these external problems that occupied me continuously I was forced to neglect my spiritual-meditative work, which had given me inner energy since my adolescence. For one month I was completely unable to spend time on inner contemplation. I felt cut off from the spiritual world and engulfed in material problems. But basically it was I myself who allowed myself be submerged to the extent no comforting spiritual word could reach me any more. It was clear my circumstances had to change.

Catching my breath

Shortly before my forty-ninth birthday I became seriously ill for the first time in my life. This development seems quite obvious to me in retrospect, after seven years of establishing a venture with rarely sufficient sleep. All the years I had worked out of a feeling that I needed to give the Sekem initiative enough of my excessive energy. Now I realized I had limits.

Previous page: Farming scenes from the early days.

4. Prevailing

In the night of March 21, I awoke with a stabbing pain somewhere in the region of my heart and with difficulties in breathing. I was taken to hospital immediately. The president of the General Medical Council, a cardiologist, was my friend. He was called, but the examination did not reveal any acute danger. Despite this I could still hardly breathe and was dependant on oxygen. I remember that two days later, on my birthday, Margret Constantini, a kindergarten teacher, came to visit me in hospital with a group of children from the farm. They sang for me, but I was hardly able to appreciate this kind gesture. Hans Werner heard about my illness and recommended that I should be transferred to the clinic in Öschelbronn near Pforzheim in Germany. I only managed the journey from the hospital to the airport with the help of oxygen. I felt so ill in the air that I did not think I would survive the flight. Elfriede picked me up in Frankfurt with an ambulance, which could only travel at forty kilometres per hour for the entire two hundred kilometres to Öschelbronn, as I came close to fainting whenever it drove faster. Because of the loving care I received at the clinic in Öschelbronn, particularly Elke von Laue's curative eurythmy, I slowly regained my former state of health. After three weeks I was able to take my first steps at Elfriede Werner's side. Every day I managed to walk one step further. Gradually I was able to forget all the past difficulties which had caused such terrible pressure on my soul. Between the end of March until June I began to recover. I spent a week in the Black Forest and learnt to live and breathe again.

Then I received a phone call from Graz. An old friend in medicinal research urgently asked for help and advice. I asked my friends Elfriede and Hans and told them: 'See, I can dance again, let me fly!' Eventually they agreed. So I flew from Stuttgart to Graz via Vienna. But during the first flight I suffered another heart attack and was taken to the intensive care ward of the nearest hospital in Vienna immediately on arrival. I could hardly speak when I awoke, but I let a doctor I knew in Vienna know about my condition, and he came to look after me at once. Nobody else knew where I was. The tests showed I had a heart thrombosis; according to the doctors I needed an operation immediately, or at least a catheter examination. But I refused both of these options and only wanted to lie still and be looked after. Now I was seriously standing at the edge. My whole lifestyle would have to change if I wanted to remain alive. I would never be able to work again in the same way. I felt terribly weak. I started to take inner leave of Sekem, my family, my friends, everything. After three days my Viennese friends managed to get hold of Elfriede and Hans in Öschelbronn. Both of them flew to Vienna immediately. They supported my decision to refuse the operation. Hans looked after me with special medicines. When I was able to travel again I returned to the clinic in Öschelbronn in a specially reserved train compartment. My recovery began anew.

Once I had recovered a bit, Elfriede and Hans introduced me to Klaus Fintelmann, a pedagogy professor. He was a co-founder of the Hibernia School in the Ruhr area, a school known for its special vocational concept in the upper school. A few years later we developed the concept of the Sekem School together.

After six months of recuperation my friends took me back to Sekem. Helmy had taken over my duties with Gudrun and Mona's close help, and had grown into the task. I told them about my illness and the experiences involved in a meeting with all the staff. Afterwards an Egyptian employee jumped up from his seat and spontaneously hugged Hans, thanking him in the name of all the other workers for restoring 'their doctor.'

After my illness we decided to re-organize the entire initiative and lay a new foundation stone. The ceremony of laying the foundation stone was accompanied by music and recitations from the Koran. Everyone present and involved in the project signed the foundation stone document and then the foundation stone was lowered into the central room of the Round House. It was all very festive. Everyone was aware of the importance of this moment.

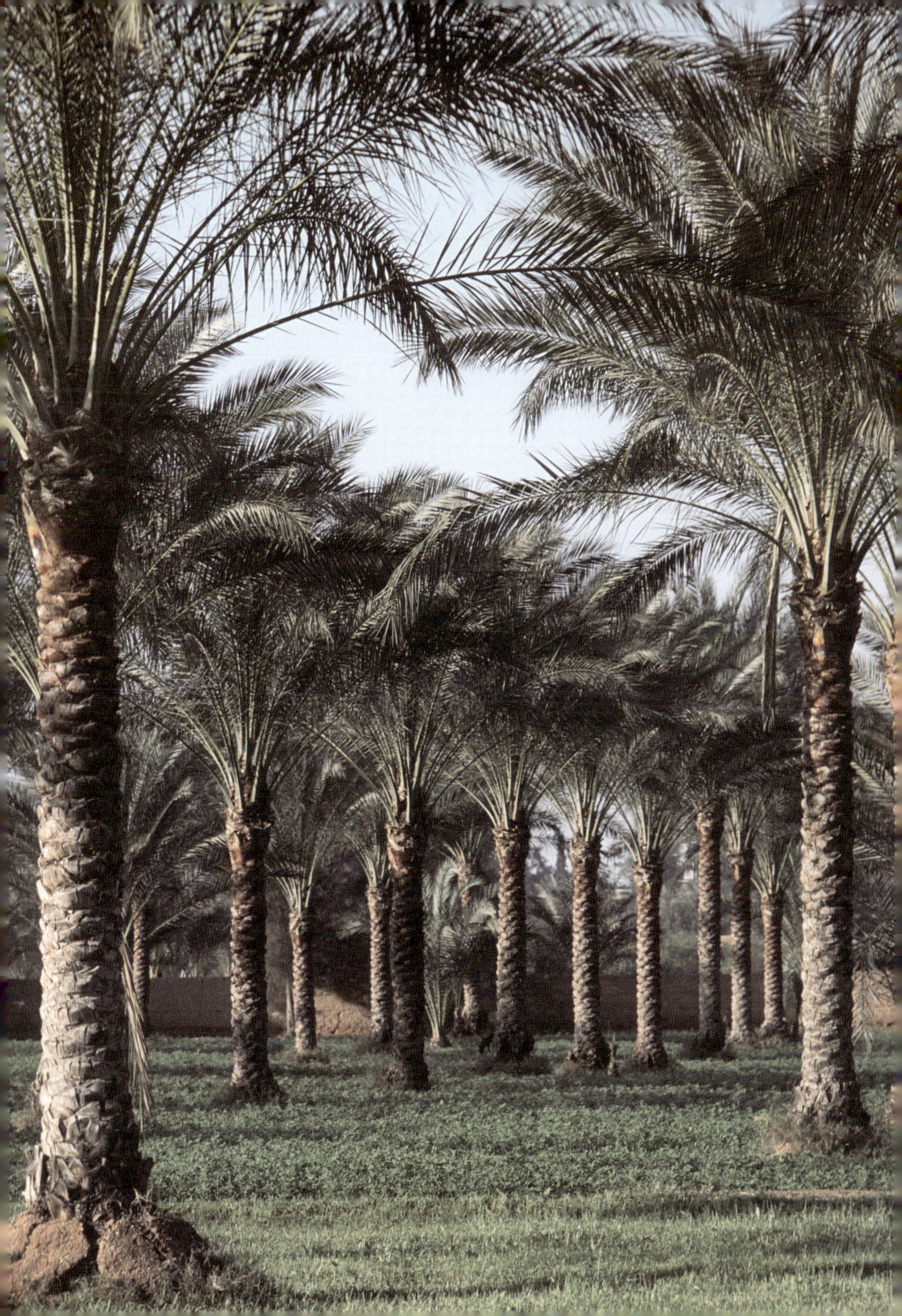

Part 3

Sekem —
A Community Model for the Twenty-First Century

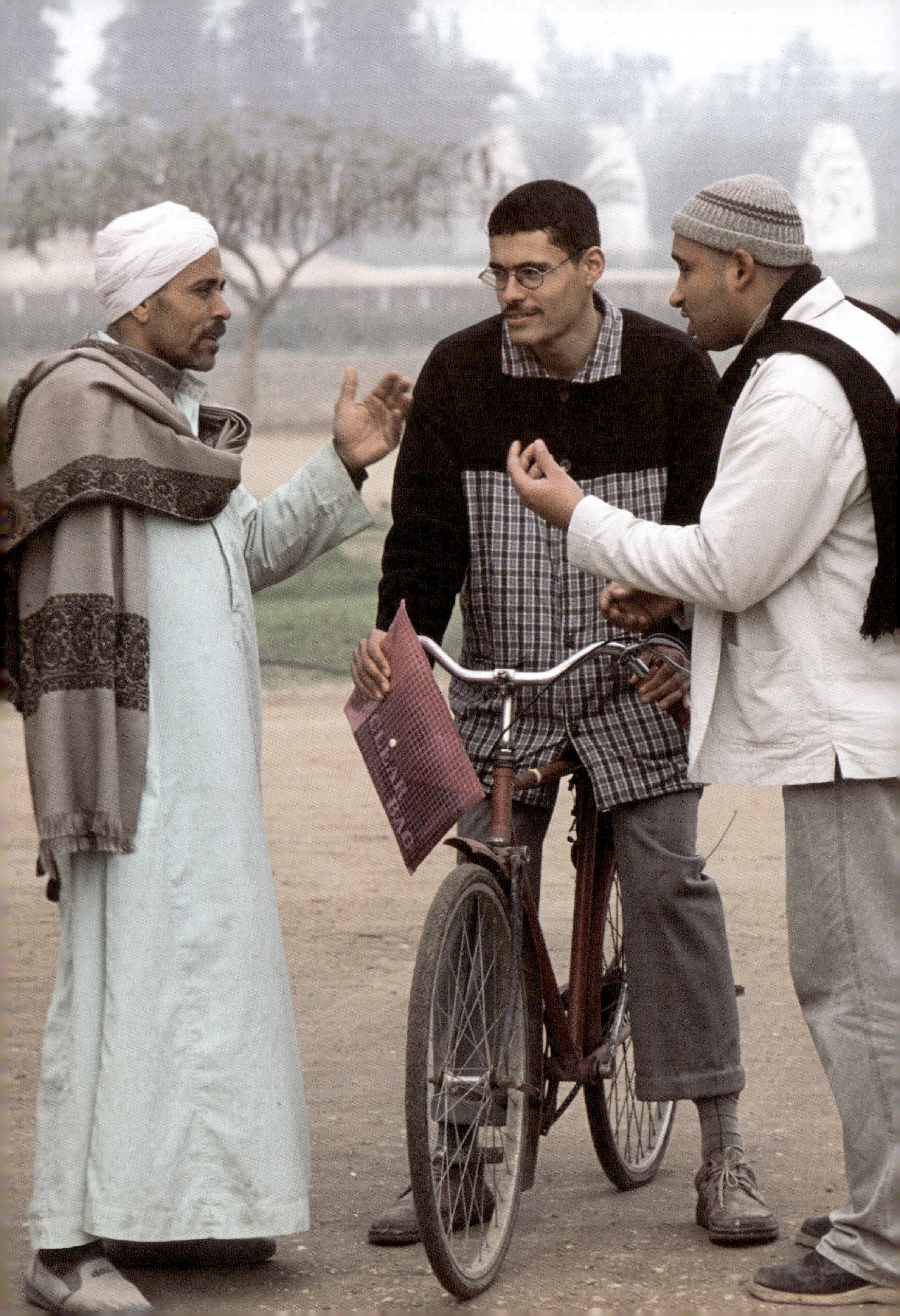

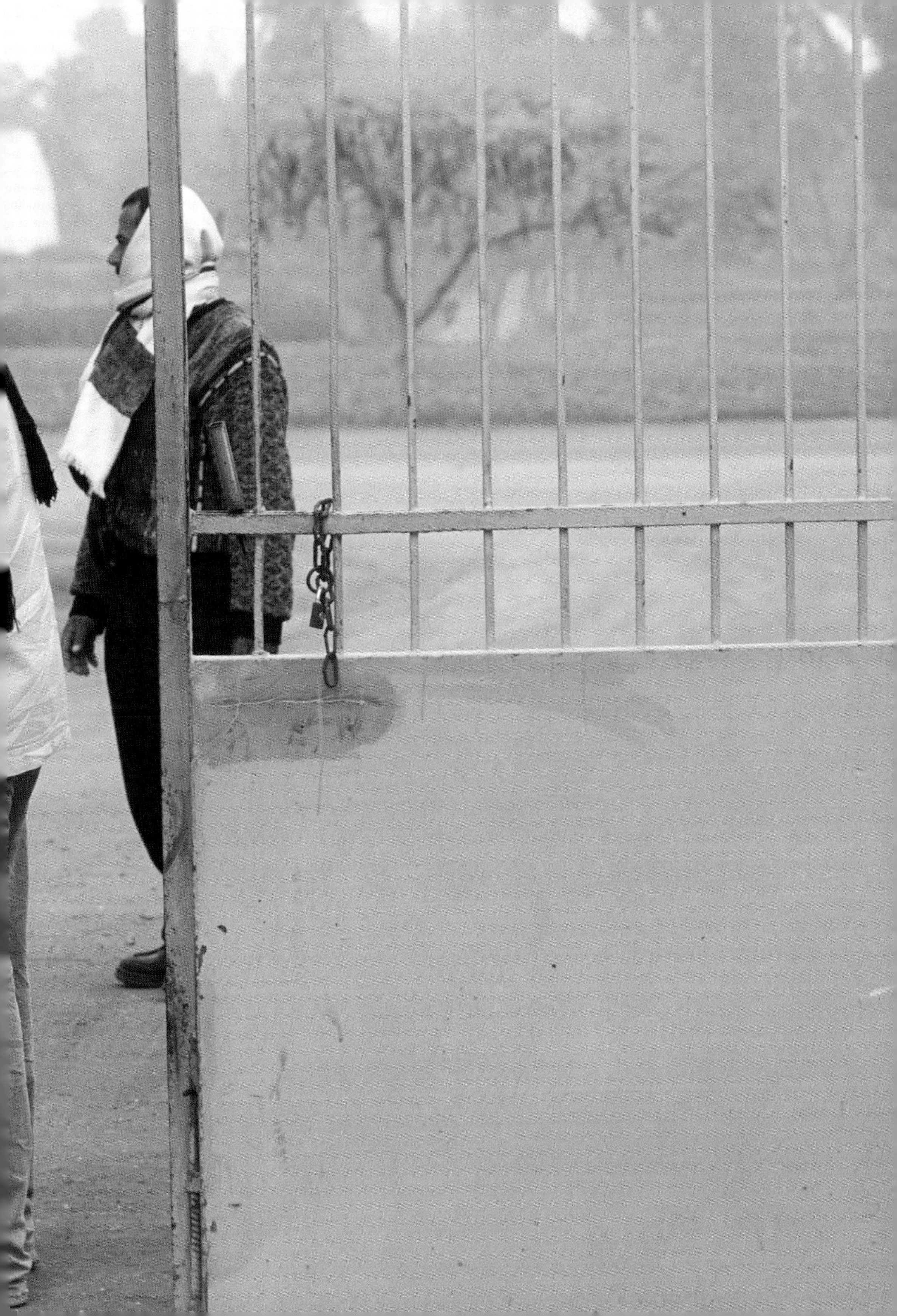

Opposite: Herbal teas being packaged.

5. Economic Foundations

Association between Egypt and Europe

Once a month all farmers working together with Sekem meet in the hall of the Sekem Academy. Every time it is very impressive when around two hundred tall, strong men with huge beards wearing long *galabias* stand up and express, often with tears in their eyes, how much they feel supported by Sekem. Their simple words which come straight from their hearts show they see an ideal of economic

Harvest time of Calendula.

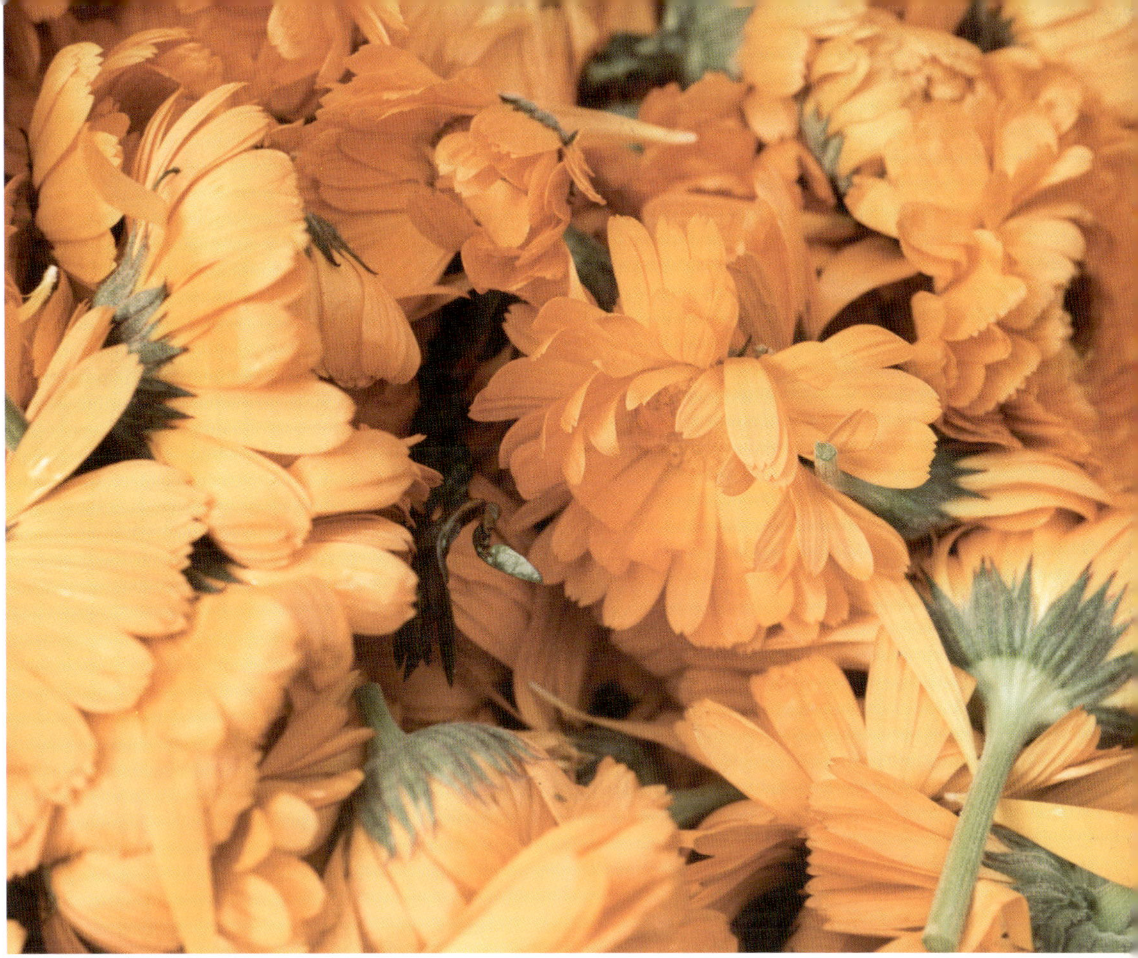

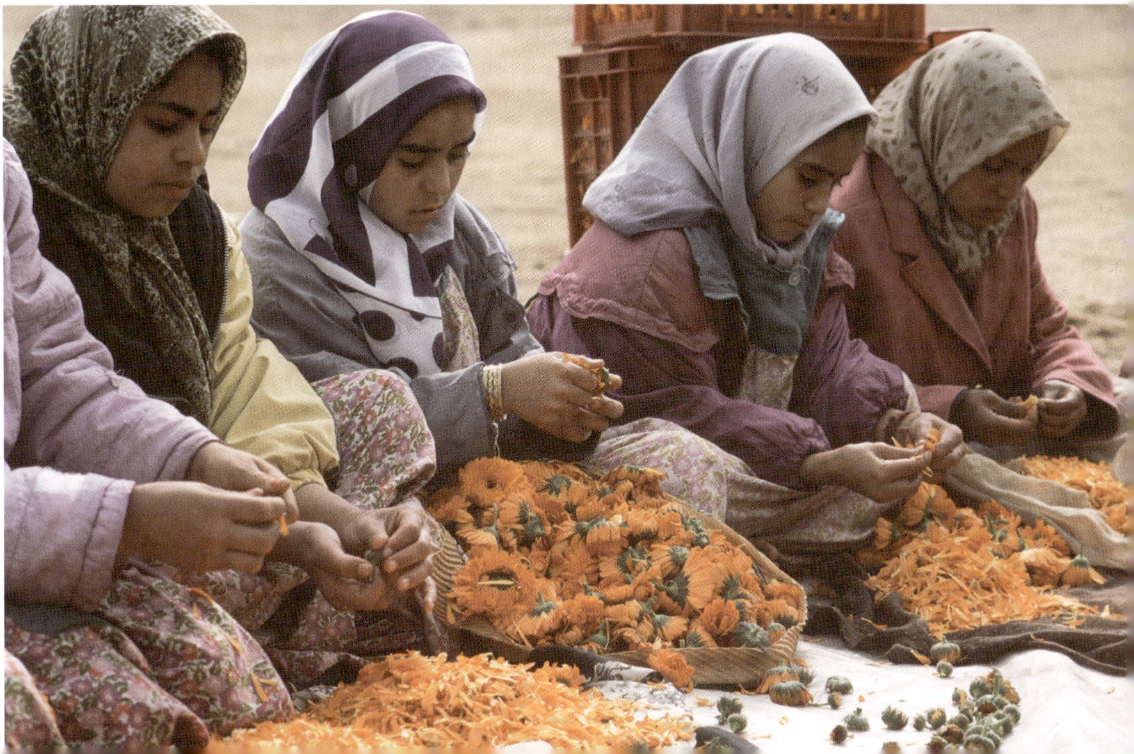

life realized, which is based on brotherliness and not competition and egoism.

Idealism and good will are not sufficient for an economic approach with these ideals. When travelling through Italy visiting biodynamic farms with Georg Merckens before coming to Egypt, I soon noticed that one of the most important prerequisites for successful economics was completely lacking: the awareness for associations. How does an association work? The products are passed through a chain of people involved in the economic process: starting with the producer (for example the farmer), over the distributor (wholesaler, retailer) and ending with the consumer. This process gives the product an added value. Nowadays, generally nobody knows about the life and work conditions of the partners involved in this chain. This anonymity causes people to think solely of their own advantage and to try to get the best deal for themselves. At every point of the chain there is pressure on the price. And the consumer, as the final link in the chain, usually also pushes the price down by choosing the cheapest product because of ignorance of the production process.

I imagine an association where the whole value-adding chain becomes transparent. It has to start with the consumer. He is asked which product he wants, what its quality should be, and how much he is willing to pay for it. The distributors agree on a percentage of this known final price for themselves. Finally the producer is given a price, and he knows the conditions which have made it. Everyone involved in the chain is under the obligation to keep to the price they have agreed on, and to deliver the product to the consumer in the state he desired it. An association is founded on an agreement that gives security to all involved. The basis of an association is thus mutual trust or, in other words: economy based on fraternity. Everyone involved in the economic process is aware of the others and recognizes their mutual dependence.

Sekem's association chains developed step by step and always through personal contacts. Here I need to mention our friendly partnership with Roland Schaette, who worked with us from the beginning to establish the associations. He also repeatedly gave us inspirations for further pharmacological products.

After our business extracting the active ingredients of *Ammi majus* ended when the American company was given up, we started planting medicinal herbs in 1983, which were initially used for teas. I wanted to expand this business into a much larger venture. At this point I met Ulrich Walter, who was just founding his business Lebensbaum for ecological products in Germany and who was looking for good quality medicinal herbs. With the support of the GLS Gemeinschaftsbank in Bochum it was possible to establish a completely transparent association chain. Ulrich Walter often visited Egypt and kept himself informed about the working conditions and production process here. With the growing consumer demand we started looking for further land for farmers to grow medicinal herbs. We had to increase the supply to suit the demand. This in turn led to the expansion of the Lebensbaum company.

I became acquainted with Volkert Engelsmann (the founder of the business EOSTA) from Holland via our fresh vegetable company — I will describe its founding in more detail later — who set up an association with Sekem for selling fresh vegetables. And our first partner for our cotton clothing business was Heinz Hess, whom I visited in Frankfurt and who encouraged us to start this new economic direction. Through Winfried Reindl I got to know Dr Götz Rehn, the founder of the organic food chain Alnatura, who started a partnership with us. Peter Segger from England also has to be mentioned in this context, who introduced Sekem to the English market through his business Organic Farm Foods. Ulrich Walter was one of our first association partners with his business Lebensbaum and he knew Bart Koolhoven. Together with Sekem they founded the company Euroherb. Recently Thomas Hartung (founder of the business Aarstderne) from Denmark has joined us. Helmy was soon the expert in charge of coordinating the associations. He founded the IAP, International Association for Partnership, an association where the partners can depend on each other. All the companies profit enormously from this trust and have been able to expand their businesses. Sometimes I wonder whether it was just good luck that I met all these splendid, inspired people, or whether heaven helps people who strive towards an ideal. All the people involved with the association have held together despite the great difficulties and hardships that occur when attempting to tread new paths in a hostile environment. But we were always able to discuss arising problems with great openness.

A visitor from Cyprus

I was sitting in my office in Cairo one day when a man from the Greek part of Cyprus introduced himself to my secretary. Shortly afterwards a lively, active businessman entered and told me about a project that

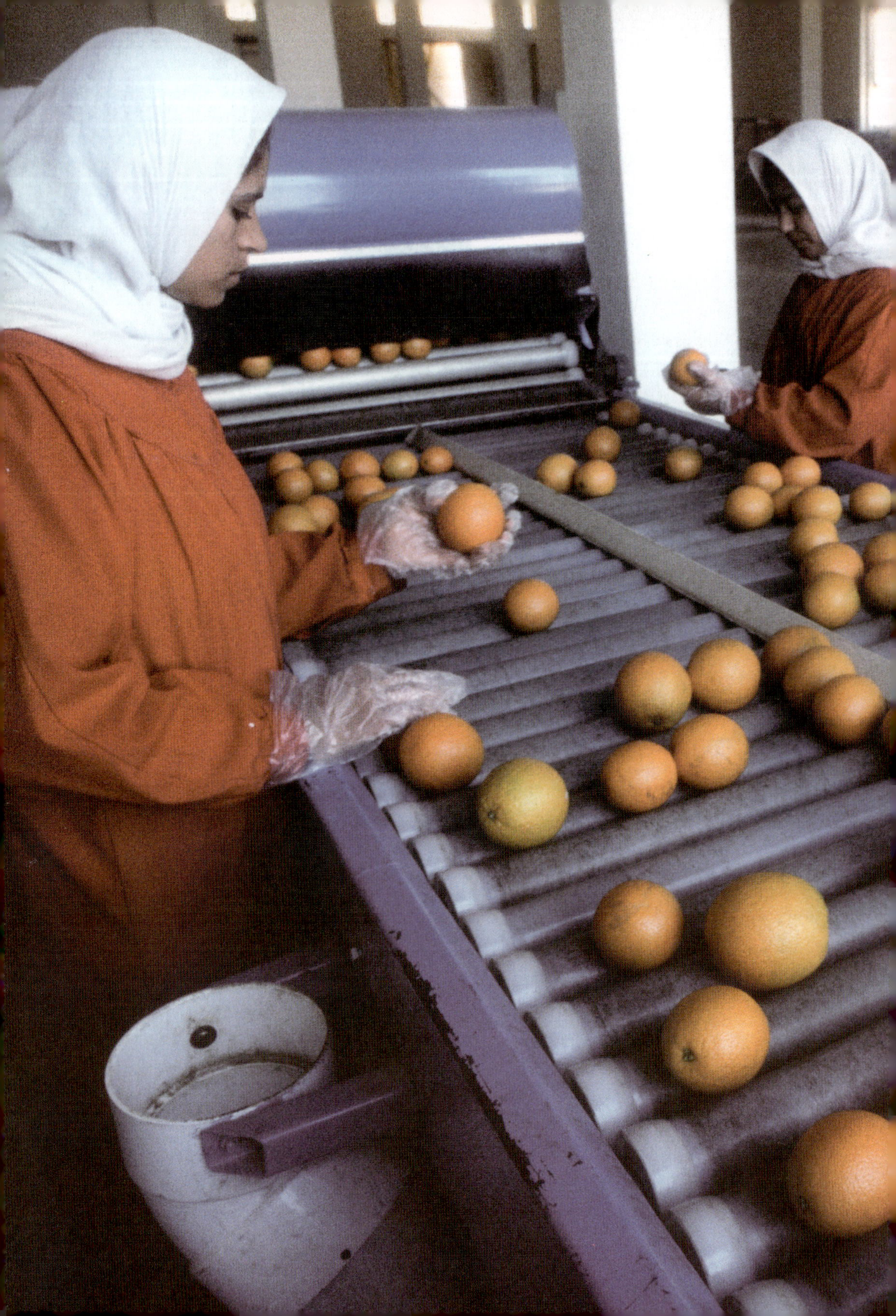

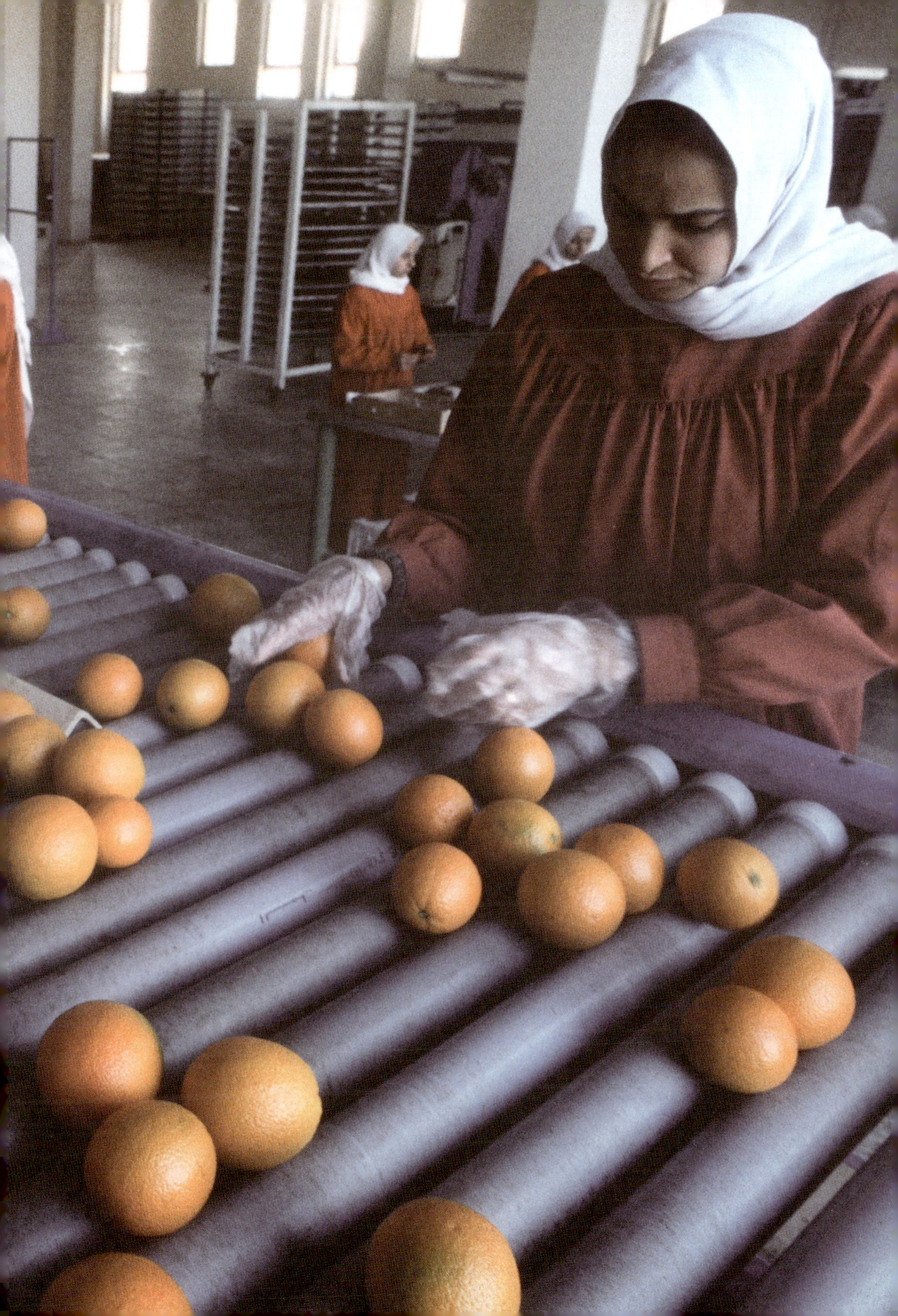

he had set up in my birthplace Mashtul in Egypt. 'What have you established, Mr Takis?'

'I have transformed vast areas of land into a vegetable producing venture, built a packing house and bought refrigerated vehicles which deliver the fresh produce to the airport, from whence they are flown to England.'

'Very good, Mr Takis! And is there a problem?'

'All the Egyptian banks have advised me to enter into a partnership with Sekem.'

I asked him: 'Why?'

And he answered: 'Because we people from Cyprus do not know how to deal with the way Egyptians work, and have suffered financial losses for years because of it!'

Up until that point he had tried to run his business exclusively with workers from Cyprus. I thought about it: so far Sekem had only produced fresh foods for its own use. Should we start trading with fresh produce? While listening to him I realized he had established what I had always wanted — to sell fresh produce. Finally I asked him: 'How do you cultivate the vegetables?'

'With artificial fertilizers and pesticides, of course.'

'Where do you get your seeds from?'

'They are hybrid seeds from England.'

Now two souls were struggling in my breast: on the one hand this man had experience in marketing fresh produce, on the other hand I objected to the chemical methods he used. I made a quick decision — for him! His experience was very valuable, and everything else could be tackled later. Helmy travelled to Mashtul to look around Mr Takis' business. He was horrified when he came back.

'It's not a food business!' he exclaimed, 'It's just artificial fertilizers and pesticides.'

I replied: 'Then we'll just have to transform it into an organic farm!'

Together we founded the Libra Company in which Sekem had a fifty percent share. We learnt a lot of valuable information about logistics and customer care from our partner. Mr Takis often came to visit Sekem and we showed him the biodynamic way of farming and its effects on the health of humans and the earth. Basically he also realized the damage conventional farming did to the earth and the plants; but at the same time the businessman in him saw his profits

Previous page: Sorting oranges in Hator Company.

and money. In his opinion organic farming made the products prohibitively expensive.

I asked Mr Takis to travel to England to find out about the market for organic produce there. At first he refused, but eventually he was persuaded, although he returned without much success. In the meantime I met up with Volkert Engelsmann, our Dutch business partner, and asked him: 'What would you think if we starting producing fresh organic vegetables?' He answered: 'That would be great!' So I asked Georg Merckens, the expert in biodynamic farming, to come and visit us and we discussed how to establish a business with fresh organic vegetables. Then we started cultivating vegetables on the other farms meanwhile belonging to Sekem.

Despite all our previous learning and observing this enterprise turned into a costly venture. It started with the fact that it was difficult to obtain seeds for the kind of vegetables the customers wanted. Then the yield was about one half of that calculated because of adjustments that had to be made. We also had to perform many insect inspections! Added to all that a sand storm raged over the farm for a few days, tearing the greenhouses apart and destroying all the work we had invested.

During this time Helmy travelled all over the country giving advice to the farmers. Our deficits increased, just because we had decided to do business with fresh organic vegetables without sufficient farming experience. But we wanted to set an example for Egypt, to prove that it is possible to produce organic food here. Every time something went wrong, or we looked at the figures, we clapped our hands together in a friendly way and chanted: 'We will manage! We can continue and will not give up.' Sometimes we would say as a joke: 'If only we had a factory making screws. We could be millionaires by now with the amount of time and energy we've invested in this project!' We remained certain throughout: with that amount of commitment the good spirits could not abandon us.

We founded a new company for the fresh food enterprise: Hator. This branch of our venture, we realized from previous experience, would need a logistics genius to manage it with the ability to assert himself. This person would have to make sure that the produce was delivered to the farm from the fields by a certain time, so it could be cleaned and packed by the deadline. At the same time the necessary customs documents had to be presented to ensure that the produce would catch the ships and aeroplanes to Europe as planned — or alternatively be delivered daily to the Egyptian grocers. The coordination required had to be performed with military precision to avoid great financial losses caused by spoiled food. Finally Gudrun started managing Hator, as she had experience with novel and challenging tasks. She taught the employees, about seventy young girls, with untiring commitment and dedication. Her training courses were held in the Mahad, our centre for adult education founded in 1987, where she taught the hygienic measures necessary for dealing with food, starting with washing hands to wearing gloves and using special protective clothing and hats. She checked the quality of the vegetables the farmers delivered, and made sure they were cooled correctly. She also ensured all the necessary processes were performed in swift sequence.

Eventually we ended our partnership with Mr Takis in mutual agreement, as he wanted to follow his own business. We were thankful to have learnt about the requirements of marketing fresh produce from him and still remain in friendly contact.

Libra has to survive

Sekem leased plots of land throughout Egypt for cultivating its medicinal herbs. An advertising campaign on the television and in the local media reporting on environmental and health issues made the firms Sekem and Isis and their tea products known throughout the country. Following this broadcast many people asked us about biodynamic farming methods, among them Chaled Abuchatwa, a large-scale landowner and mechanic from the north of Egypt. He was so enthusiastic about the idea that he promptly started cultivating organic medicinal herbs on his lands. He came from a well-known family and was greatly respected. His neighbours would peep over his fence and ask: 'What's happening here? No artificial fertilizers? No pesticides? It smells lovely.'

With the increasing interest in biodynamic farming we saw the necessity to expand our farming and marketing to products beyond the medicinal herbs we cultivated. To accommodate this increase in cultivation Sekem leased further plots of land throughout Egypt and employed seventy new farmers. Farming has its own rhythms and thus a specific turnover pattern: for one season of nine to ten months it is necessary to invest continuously — the time during which the product grows, is harvested, dried and processed — and only then a 'product' is produced which brings in money. This agricultural products business was taken over by Libra.

5. Economic Foundations

During his time, one of the most significant cotton dealers in Egypt had been my grandfather on my mother's side. His right-hand man, who administered the huge cotton warehouses, was called Abdullah Lefef. One day a young farmer called Ahmed Rashad appeared and introduced himself as the grandson of Abdullah. I employed him for Libra on the spot. Ahmed become involved with the new tasks with admirable calm and administered the difficult cultivation of agricultural products. He worked together closely with Georg Merckens and learnt the basics of biodynamic farming while travelling around Egypt with him. For some time we made a wonderful threesome training farmers to convert their farms to the biodynamic method. Georg Merckens had gathered a lot of previous experience during his advisory work in Sicily and Mexico. He talked to the farmers while I translated, adding verses from the Koran to open up the souls of these people to the new ideas of biodynamic farming. And Ahmed, as an Egyptian farmer like them, was the socially uniting component. 'We are not only called upon by Allah to care for the earth, which He has bestowed on us, but also to heal what has been destroyed,' I said. When plants are subjected to artificial fertilizers during their growth they absorb much more water into their fruits, but produce less vitamins, which we need for our nutrition. The Koran differentiates between two food qualities: food which people are allowed to eat — it is called halal — and exquisite foods, which are called *tajeb*. This is explained in the Koran in Sura 16.114 'The Bee.' How does food become *tajeb*, exquisite? With the help of many trace elements in the ground the plant develops its active ingredients. Depending on the kind of plant, micro-organisms help to prepare the ground around the root hairs so that these trace elements become available for the plant to absorb. Artificial fertilizers intervene with this living, sensitive process and change it in two ways: they change the pH-value of the ground and destroy the micro-organisms. The result is that the plants absorb more water and fewer trace elements. This leads to larger fruits, but they are not *tajeb*, exquisite. If we do not look after the living energy of the earth by using compost preparations, people cannot be nourished properly, which has far-reaching consequences, even for their soul life.

I went on explaining how people had sunk ever deeper into materialism during the last century. Scientists emphasized the importance of nitrogen, potassium and phosphorus for growing plants and showed how the yield could be improved by adding these substances. But they overlooked the fact that the amount of artificial fertilizer had to be increased continuously to achieve the same results and that the plants became more susceptible to pests, which had to be controlled with poisonous chemicals. Herbicides were then used to keep the weeds in check which grew simultaneously. The consequences were soil erosion, compacting of the ground and the disappearance of whole species of plants.

Biodynamic farming heals from the earth upwards. Using compost and preparations strengthens the life energy of the plants so that they are able to grow of their own accord and not through the addition of artificial fertilizers. It also has an effect on the health of humans and animals.

After these introductory words Georg Merckens explained the practicalities of making compost and owning animals. All the farmers then received further weekly schooling on the Sekem mother farm in the Mahad.

Despite the extensive training the transition to biodynamic farming was difficult for every single farm. We had to reckon with at least two or three years of high crop failure rate, as the farmers had so many new things to learn. The products also turn out differently in the first years of omitting artificial fertilizers and are more susceptible to pests until the new equilibrium has been established. The practical implementation of organic farming also requires a personal cultivation of awareness. This way of farming depends on an alert consciousness which is able to think preventatively and in an interrelated way. But the concept of prevention is still unknown to most people in Egypt.

When I looked over the balance sheets of Libra with Helmy and Ahmed and we saw the deficits despite the intense training and supervision of the employees, we would say to each other: well, who is willing to invest in ideals nowadays! Somebody has to set an example during the transition phase towards implementing biodynamic farming in Egypt, to prove it is possible and to be willing to carry it financially. This is Libra's task! And apart from that, this work gives us the opportunity to educate people, to heal the earth, and to test the ideas of biodynamic farming in reality. After such thoughts we looked into each others eyes and decided every year anew: Libra must survive!

Ahmed was faced with many difficulties. For example, Libra grew potatoes and exported two thousand tons of them to Europe. Two to three hundred tons were rejected, because they did not conform to the standard size. This meant the product had to be withdrawn and sold at a loss. Another time we had to book space on a container ship months

5. Economic Foundations

in advance for transporting onions. The harvest was delayed and the space could not be used, but it still had to be paid. When people saw the long line of trucks filled with agricultural produce delivering to or leaving Sekem, they thought we must be the richest people on earth. What an error!

Libra built greenhouses which involved huge effort and expenditure to expand the vegetable business and be able to deliver fresh produce to Cairo and Europe even in winter. We had to buy expensive seeds, sold not per kilo but per seed, pre-germinate them in cold frames, prick them out and transplant them to the greenhouses. It sometimes happened that farmers bought cucumber seeds for 100,000 pounds which did not germinate because the seeds were either too old or 'wrong.' The greenhouses were prepared, the orders had come months ago, but nothing grew, or only grew later after new seeds had germinated. This involved high risks for our agricultural partners!

We tried to find a satisfactory solution for these people, who were trying to get used to the new way of working to the best of their abilities, and who were willing to heal the earth using organic methods. We decided to arrange a fair price for the products in advance and let Sekem or Libra carry any losses. We informed our trading partners in Cairo and Europe of the prices, which had to at least partially include these risks. This in turn made them aware about the living conditions of the Egyptian farmers and they were also able to carry a part of the transformation to biodynamic agriculture in Egypt. I wanted the price fixing to be transparent from the farmer to the consumer, which was in accordance with my understanding of 'association.' Naturally not all the business risks could be reflected in the price, as that would make the products prohibitively expensive. Again, Libra cleared the balanced.

The farmers needed investments to purchase cows, stables, compost equipment, transportation, drying devices and establishing communication systems and not least for building their own living quarters. All the costs for investments and training, for support during the conversion period and with cultivation losses meant huge financial deficits for the company Libra, as the balance could only break even if there was a profit. Despite this Libra helped and continues to help the farmers and guarantees they receive the arranged prices.

During our training meetings the farmers discovered where their products went and how the prices were devised. Helmy explained the economic relationships, while I pointed towards spiritual backgrounds.

We established the term 'economy of love.' We consciously want our agricultural business to be based on the principles of love, that is, a responsibility towards the earth, the plants and animals and to create trust amongst the people. Even when the market price was much lower than the price agreed upon, we always bought the produce of the farmers at the previously agreed price.

Added to the factors already mentioned are Egypt's expensive land prices. Egypt only has a thin strip of fertile land on either side of the Nile and the rest is desert, which is expensive to cultivate because irrigation costs have to be included in the prices. At the end of the year Helmy, Ahmed and I agreed we had succeeded in proving that arid desert country can be cultivated using biodynamic farming methods, and that people had been educated and trained in the process. We felt these results weighed favourably against the losses, and were content it was not a hopeless business. Eventually the money and effort would pay off. Again, despite its deficits, it was clear to me: Libra had to live! At the end of the year we balanced Libra's losses with the profits of the other flourishing companies, or through bank loans. The bank directors never had any doubts about the importance of our work after visiting Sekem.

Competent training became ever more necessary to enable the farmers to become more attentive and able to improve the earth, to treat insect infestations early on, and to make compost using the preparations. Later, when we added cotton cultivation to our agricultural business, we founded the Egyptian Biodynamic Association as a non-profit organization with the task of training people. It also trained agricultural advisors to send around the farms. Thanks to Georg Merckens' and Helmy's input the Association now has a lot of experience with biodynamic cultivation and particularly with logistics. The farmers each pay a small contribution towards this organization, so at least some of the training costs are covered.

We were faced with the further problem that organic products have to be certified to be viable in the marketplace. We dealt with this requirement early on, and had our products inspected and certified by the Swiss Institute for Marketing Ecology (IMO). At the same time it became clear that Libra could not finance the high daily costs of the inspectors for long. This is a problem faced by many countries unable to afford expensive European inspectors. I discussed the situation with the IMO inspector responsible for Sekem and asked him: 'Would it not be possible to have an Egyptian branch of certification?' He admitted he had been wondering how we could pay for the high costs, and encouraged us to branch out on our own. Helmy contacted the Demeter association in Germany and by 1990 we had managed to found the Centre of Organic Agriculture in Egypt (COAE), an independent company which now inspects farms in the whole of Egypt, Iran and Sudan. We found Dr Yusri Hashim to conduct the inspections. The certification branch office was based in Cairo. Klaus Merckens (Georg's son) learnt the necessary inspection and certification guidelines from the Swiss IMO and was invaluable in establishing the COAE. Klaus Merckens had already decided to move to Egypt permanently by this time. Now his work contributing to the establishment of the COAE has finished, Klaus is our fundraiser, and he represents the Sekem idea throughout the world. His wife Johanna works in the textile industry Conytex as an export specialist. Originally she trained as a kindergarten teacher and first had to acquire the basics of correspondence for her new task. But it is always a special experience to see how people can take over an unfamiliar task and grow into their work.

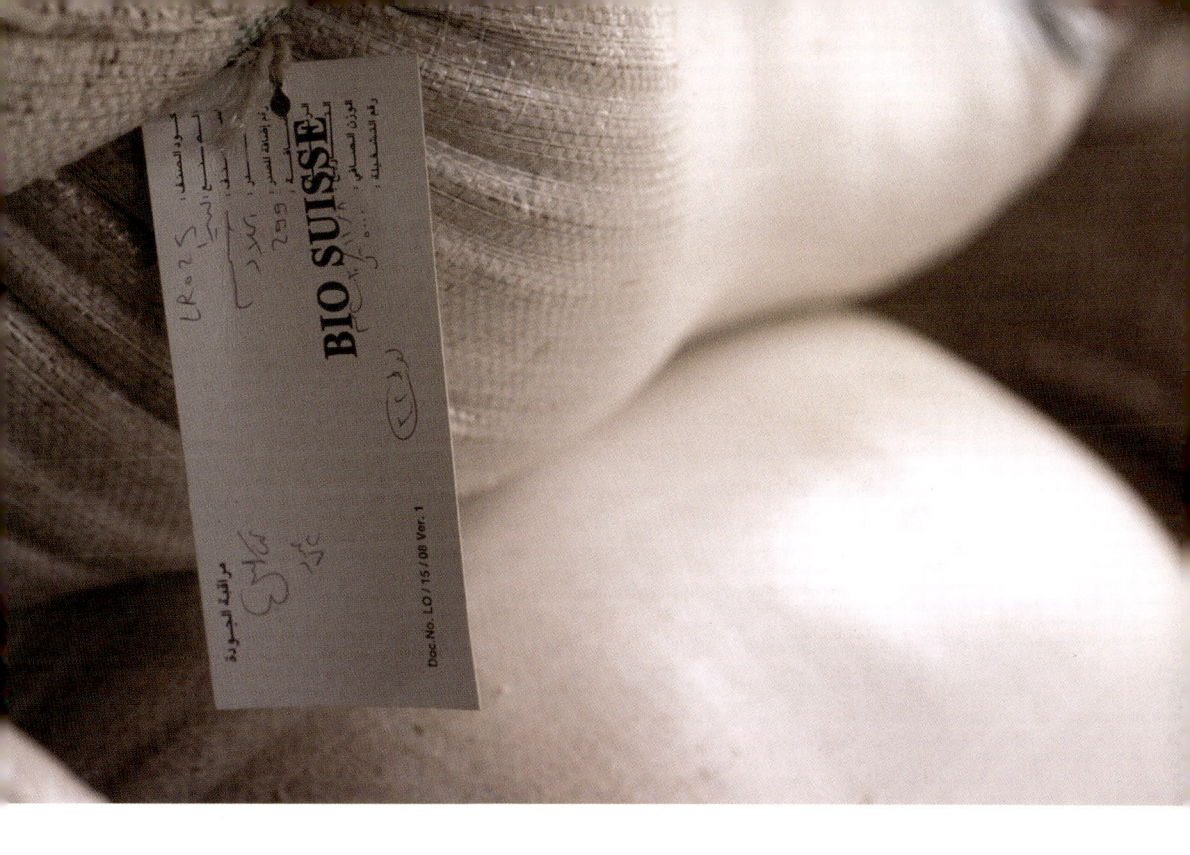

    For a long time we wondered how to balance Libra financially. Finally we decided to build a mill, bought an oil press, which could also make olive oil, and drying units. Hans Spielberger, an experienced businessman in this area, supported us with his years of experience. The agricultural products became more valuable because of their refinement, which meant a profit for the company. We hoped that this would lead to positive figures in the accounts of the following years. Nowadays the company Isis successfully sells many processed foods like honey and jam, syrup, oils and dry foods like cereals and rice, nuts, dates and figs.

We frequently received help from people in Egypt during the establishment phase of cultivating organic food. Mohammed Gaballah, an educated farmer from Fayum, heard about the idea of organic farming and was very keen. He himself only owned a small patch of land. What did he do? He talked to everyone living around his land, relatives, friends and neighbours and founded a cooperative. Three hundred people pooled their land and created a 'country without limits' where they started cultivating biodynamic medicinal herbs. They picked and dried the products themselves. We bought the herbs off Mohammed and he divided the money fairly among his neighbours. For some years this cooperative has been working with great success, which seems more than a miracle in this country. I see this Mohammed Gaballah as a hero, a man who has managed to build up a real economic community through selfless effort, the 'farmers without limits' as I call them. I often took people to visit him, to show what an inspired person can achieve with enough enthusiasm for the idea.

Chaled Abuchatwa, the large-scale landowner from the north, advertised and circulated our concept with great enthusiasm at farmers' conferences, and with his charismatic personality and power of speech stood up for Libra. Sekem also gradually became a showpiece for state visits in Egypt. Many political representatives from around the world have got to know our farm. The union of African farmers visits once a year for a week and shows interest in the biodynamic way of farming, the humane pricing system and our purchase guarantee.

Cotton — poisonous white gold

One day pesticide tests performed on our medicinal plants showed traces of residues. We were rightly outraged. Where did these pesticides come from? We were certainly not using them. After excluding a whole range of possibilities we finally realsd that they were sprayed onto our fields by the crop dusting planes applying pesticides to the

First International Organic Cotton Conference in Cairo, 1993.

neighbouring cotton fields up to twenty times during their growth period.

Once I realized this I complained to the minister of agriculture. 'We want to cultivate organic produce on our farms without using poisons,' I said, 'And you are destroying our efforts. We are powerless against crop dusting!'

He looked at me with astonishment and asked: 'What do you want me to do, is there an alternative?'

'Stop spraying the pesticides!' I said.

'Do you know what will happen if we do that?' he asked me. Only then I realized that this man was in a difficult position with regard to the chemical companies.

I discussed the problem with Helmy and Georg Merckens and asked the latter whether he knew of an organic method of protecting the cotton plants. He advised us to study the insects that harmed the plants and to learn their way of life. We asked an entomologist to explain the behaviour of the insects in question and to show us studies of their developmental stages. Then we asked different scientists how we could stop these insects multiplying using organic methods. The Egyptian scientists Dr El Araby and Dr Abdel Saher helped us and started examining the test fields prepared for this purpose. They

Helmy during the cotton harvest.

found that with the increasing heat and plant growth, small insects like aphids, physapodes and white fly developed. They can be caught with sticky yellow panels, as the yellow colour attracts them. While the leaves are growing they are susceptible to the leaf caterpillar spodoptera, and later follow the pink bollworm, pectinophora, to mention only the most damaging pests. During the hot summer it is possible for four generations to breed, and they remain a danger to the plants until the harvest. It was only too understandable that twenty or more crop dustings were used against these pests.

Dr Youssef Afifi, an entomologist from the University of Cairo took this chance to apply his laboratory experience with the spodoptera butterfly to large-scale field cultivation. Nobody knew if it would work, but we started cultivating a nine hectare field in the Nile delta. We spread simple cone traps with pheromones over the field before the butterflies first appeared. This caught the butterflies before they could lay their eggs, which would otherwise hatch into rapidly multiplying, leaf-eating caterpillars. Then we had to protect the cotton capsules, in which the seeds surrounded by fibres — the actual cotton — develop, from the pink bollworm, which could endanger the entire crop. Dr Afifi used tubes which spread a captivating fragrance. When the pink bollworm smelt it they became so confused that they could not find the capsules. In our first year the confusing effect did not last long enough and eleven percent of the capsules were damaged. We

5. Economic Foundations

soon corrected this and within a short time the damage was less than that achieved by chemical crop dusting in conventional cultivation. Once the first harvest had been weighed we found there was a ten percent higher yield of raw cotton than the average of the area. This was a result to be proud of and we attributed it to the earth enlivening and plant growth enhancing methods of biodynamic farming.

Once we thought we had solved the problem, and that crop dusting pesticides over the fields was superfluous, we sent out invitations to the first international organic cotton conference in the world in Cairo. About 120 specialists came together. As part of the conference they were able to visit the nearest of the nineteen biodynamically farmed cotton fields during the harvesting process. Egyptian television also attended and broadcasted a very positive report — people greatly admired our success. The agricultural minister had followed our progress with interest and arrived at the conference with his staff following our invitation. In his speech he said something like: 'You have my great admiration for your efforts. But who knows if you can achieve such success again! First you will need to prove your results more than once!' So there was nothing we could do but to continue testing our developed methods of controlling pests. Every year the minister chose some of the most polluted areas according to a plan and told us if our method succeeded there, then he could make his decision. I thought he was acting as a responsible person.

The testing fields were spread out throughout the whole of Egypt and Helmy spent all his time travelling. The fields had to be supervised day and night, and if quick action became necessary he had to be on site. But Helmy's efforts alone would not have sufficed if his wife Konstanze, whom I greatly valued, had not been willing to support his tasks. Because of her upbringing she was used to regarding leisure time as something important. But here she had to experience the opposite. She and her four children had to live without Helmy for long periods of time and often he would only come home exhausted late at night.

After three years we had finished testing and were able to present the results. The minister kept his word and reacted with courage, ordering the end of crop dusting planes applying pesticides to the fields. First an area of 200,000 hectares was cultivated completely without pesticides, then, one year later, this area was expanded to 400,000 hectares which incorporated the entire extent of cotton cultivation in Egypt. Organic methods of controlling the cotton plant pests were employed in the entire country.

It is hardly possible to describe the repercussions of this decision. 35,000 tons of pesticides could not be deposited on the fields by the chemical industry. The people involved had tried to fight against organic cultivation and had incited the press. We took it with equanimity and reacted calmly to anything bad. I believe the attacks we had to withstand could have destroyed our community. I will describe one particularly harsh attack later.

One of the worst chemical poisons had been abolished. Dr El Beltagy of the state agricultural research institution said in a speech that even if the United Nations had decided that Egypt should practise pesticide-free cultivation, they would not have succeeded in implementing it! And the scientists in all the universities of the country would never have come to an agreement on the matter. It was the solely effort and will-power of the Sekem community which had achieved this healing act for the country.

The damaging pesticide spraying had ended. But what about the hundreds of tons of organic cotton, which is normally subjected to the strongest chemicals during further processing? We founded Conytex, a company in association with Sekem to cover this area. After seeding, ginning and spinning, the cotton is mainly made into knitted fabric. Normally the raw material is treated with chemicals and dyed with ecologically harmful and poisonous dyes, also used to dye thread. Conytex uses mechanical rather than chemical methods, with the result that the material can shrink up to six percent. But the customers can adapt to this. The dyes are used according to Demeter guidelines and are environmentally friendly, biodegradable and skin friendly. There are repeated quality controls after each processing step. Starting with cutting of the material Sekem carries out all further processes.

At first the staff did not understand the unusual company name Conytex. But I wanted to acknowledge my dear daughter-in-law Konstanze by bringing part of her name together with the product requiring processing. 'There has never been a company named after a person before!' I was told. I calmed everybody down by pointing out that Arab people associate the syllable 'con' with cotton, which was also confirmed.

During the cotton testing period we started dealing with the German textile company Hess-Natur, and shortly after they sent us the first patterns. Again we were faced with tremendous logistic problems, coordinating all the many delivery businesses. My wife Gudrun stepped in, as she had at the start of each of the companies.

5. Economic Foundations

Above and overleaf: Workers in Conytex' sewing workshops.

Her task was to train the staff and set up the logistics. The people she trained learnt a lot. When an Egyptian makes a mistake, he likes to shrug his shoulders and say: *'Ma'alesh!* It doesn't matter!' Everyone knows Gudrun will not let them get away with this. Nobody understood and understands how to consistently train and inspire people like she does.

Johanna Merckens took over the coordination of the logistics from her, and additionally was also responsible for export. I discovered she had the pedagogical ability to mediate between the customer and the internal interests of the company, although she was actually a trained kindergarten teacher. Although in her previous life she had rarely had to write business letters or supervise the punctual delivery of the many pieces necessary for the textiles — a continuous struggle! — as well as inspect the quality of the sewn pieces, she joined in with enthusiasm and joy. Konstanze, who was an artist all round, worked as the fashion designer with the same energy she had employed when setting up the cheese dairy and music lessons in the school. She receives a task, sings and starts working! Joyfulness sparkles around her. The team was later

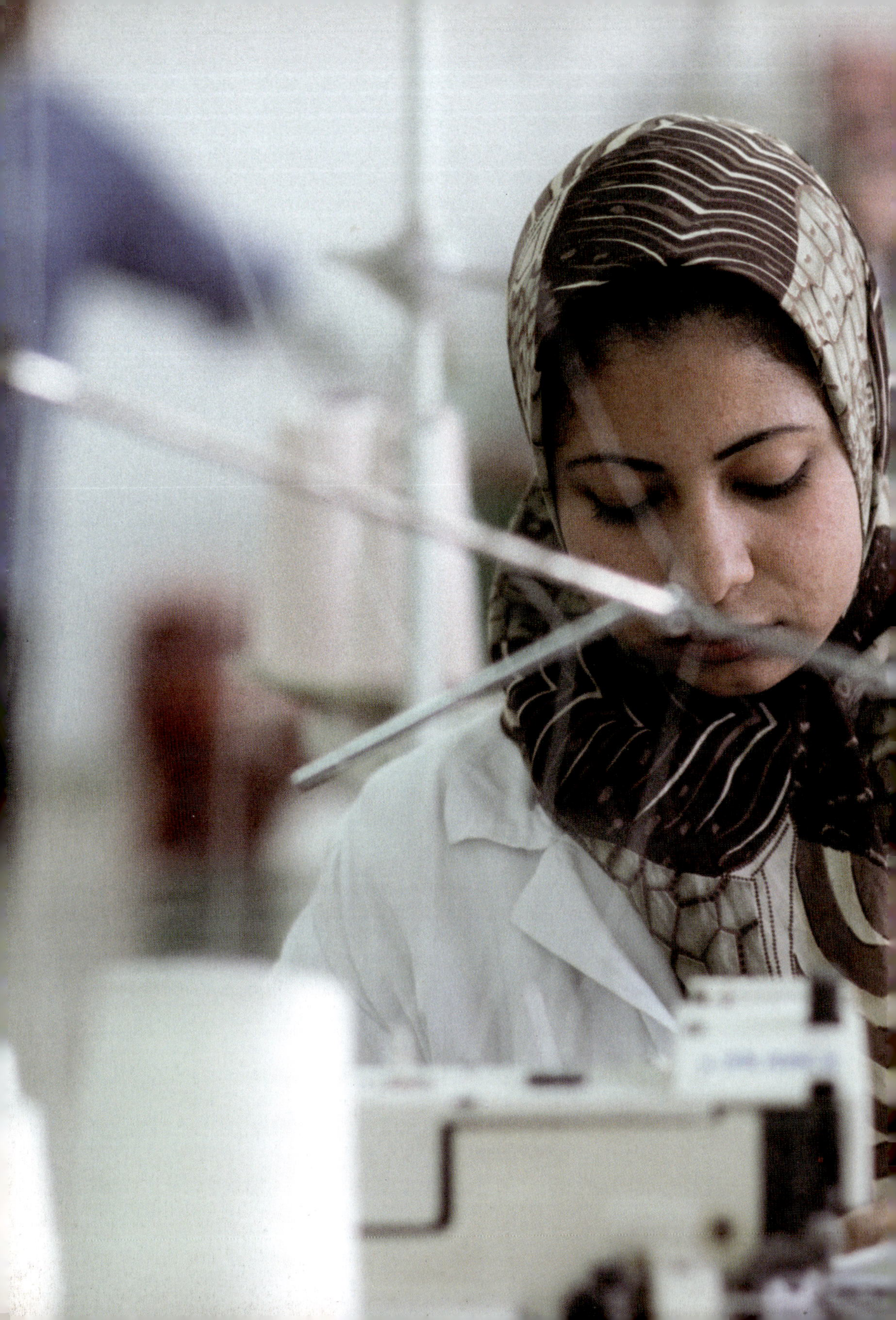

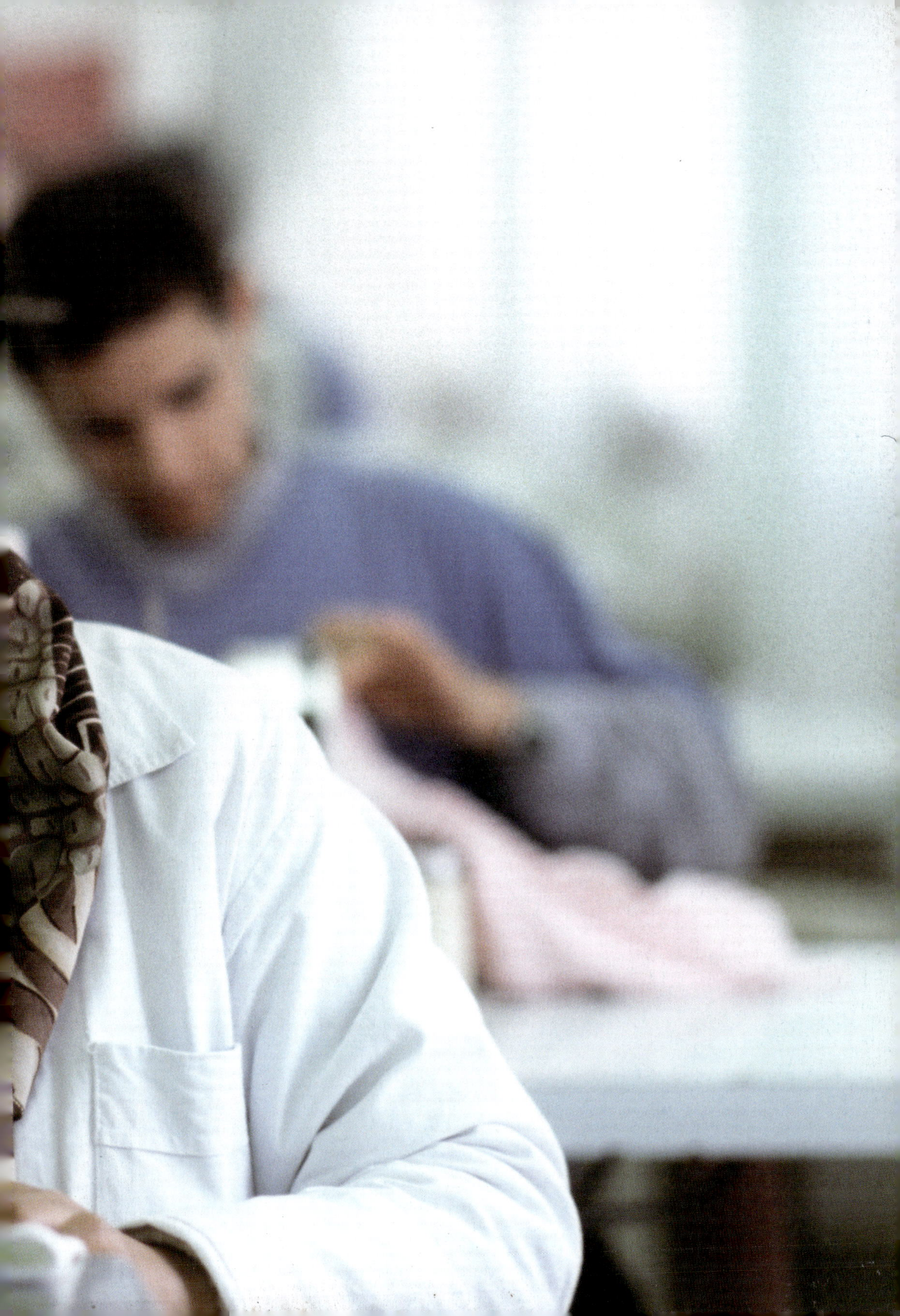

expanded by Maria Raileanu from Romania, who came to us with her two children and has worked in quality control since.

The Conytex products underwent rapid expansion. First we made natural coloured baby suits for the German market as instructed by Dr Götz Rehn of Alnatura. Because of babies' and toddlers' allergies, organic textiles for young children are very popular in Germany. Due to the continuous expansion of the company, Conytex had to move twice on the Sekem grounds. We also established a printing works for printing T-shirts or company labels. Nowadays Conytex has about 100 machines, produces about 2000 items a day, and employs about 180 workers. Because we deliver to the international market the business remains stable. We would still like to expand into the local market and export towards the east.

The 'sun-worshippers'

Before the government banned crop-dusting planes applying pesticides over the cotton fields, they had fixed contracts with the crop dusting companies and the chemical industry. Because of these the minister of agriculture could not have agreed to our demands to stop the spraying after the first year. But after three years, once we had proven a viable organic alternative on our test fields, he cancelled the contracts. This was a courageous step. Some people in the ministry were still saying that we were destroying the country. Naturally we tried to counteract this view by explaining our work. But during this time I often silently prayed that everything would turn out well!

A few weeks later articles started appearing in the large daily papers in Cairo which declared that only the rich profited from organic farming, as they were the only ones who could afford the expensive prices, which were naturally highly exaggerated. Other articles stated that not even the people of rich industrial countries could afford organic produce — and if even they could not, then the poor countries could certainly not do so. How could hundreds of millions of people in the world be fed if the crops were not improved by artificial fertilizer? Organic farming was declared to be a loser's method. We were even accused of wanting to let people starve. Sekem was mentioned by name in many of the articles and I received anonymous threatening phone calls. But there were also encouraging voices that said: 'Don't give up! You are doing good work!'

5. Economic Foundations

There was a general atmosphere of conflict throughout the whole country, and the subject became widely discussed, which could only be good in the long run. We noticed that the attacks did not influence the sales of our companies' products, even though they were supposed to damage our reputation. We were called an 'élitist company,' supposedly only catering to Germans.

We were able to cope with all the attacks until one day an extensive article appeared in the local paper with the title 'the sun-worshippers.' A journalist had visited Sekem without our knowledge and had photographed us standing in a circle on a Thursday afternoon, at our end of week assembly. He asked the question what we were doing, and then answered it himself with the statement that we were worshipping the sun! He had photographed the Round House, and mentioned other round shapes in and in front of the company buildings — according to him all symbols of the sun! Finally he cited a man from the education authority as literally saying: 'Dr Abouleish stood in front of the class and asked the children: "Who is your God?" The children truthfully answered: "Allah!" Then he told them: "No, not Allah. I am your Allah!" I experienced this myself ...' the supposed education inspector further lied.

Worshipping the sun for Muslims is like worshipping Satan for Europeans. The people were in turmoil. Something like that in their country! They expressed their indignation. Sekem workers were harassed with words like: 'Is it true? Are you sun-worshippers?' and had stones thrown at them. The article circulated throughout the whole of Egypt.

Then I got a telephone call from the head of the secret state security police, who invited me around. When I entered his office I saw the article lying on his desk. He pointed towards it, and said laughing: 'What do you say to that stuff?' Because I did not know his view I waited in silence. He continued: 'We here know that not a word of the accusations against you is true. But I advise you to defend yourself and take legal action against these people! You cannot let them get away with these accusations!' Now I had proof of what I had always assumed: Sekem also had spies from the state secret services amongst its workers like all larger companies, who were placed there because of fear of fundamentalists. I followed his advice and started a court case against the paper, knowing fine well it would take years.

Based on this article the prayer leaders in the mosques around Sekem started to stir up animosity towards us, spreading the word that we did not worship Allah, but the sun. Amongst their worshippers were Sekem workers, who knew that this was not true. But nobody would be allowed to stand up in front of all the people and say something against the imam! I began to fear that the chemical companies had won after all.

Should we fight against the animosity, or choose another way, one which was peaceful and took the wind out of the enemy's sails? — I decided on the latter course, the one I had already used with the Bedouin. Ten of my trustworthy staff members were given the task of inviting to Sekem all the people mentioned in the article, as well as the mayor and influential sheiks of the area. We fixed a date and I stressed that everyone was responsible for ensuring that the people assigned to them actually came. On the Thursday I met up with them in the Mahad. They entered, a large group of men in long flowing gowns. I welcomed them by handshake, which they returned unwillingly. But I remained calm. Once everyone was seated I asked a sheik to read a verse from the Koran, which he did with his beautiful voice. Once he had finished I beckoned Sekem musicians into the room to play a Mozart serenade. Suddenly a man jumped up furiously, banged his fist on the back of the chair and shouted: 'We will not listen to this work of the devil!' I walked up to him while the musicians continued playing bravely, and said: 'Calm down and finish listening!' After that episode all the visitors let these 'terrible' sounds wash over them. Once the musicians had left the room I invited the men to express themselves. One of them stood up and shouted: 'Music and art are forbidden in Islam. The Prophet said so!' I calmly asked back: 'Does it say so in the Koran?' — 'No, the Prophet said it!' — 'I believe every word in the Koran, and also those of the Prophet. I only need to see it first!' I answered. — 'I'll bring it to you!' — 'Good, I'll wait until you bring it!' The meeting started in this way. The atmosphere was terribly strained and threatened to escalate beyond control at any moment.

Because of the questions I started telling them that Allah had chosen the human being out of all his creations to be his successor. Some of them nodded, because I verified everything I said with verses from the Koran, quoting them by heart. Allah says: 'We are responsible for the earth, the plants and the animals.' Allah had initially given the responsibility to the heaven and the mountains, but they had refused. It was too much for them. Only the humans took it upon themselves. They all knew these verses (33.72).

Now I continued talking about the dead and living earth, as is writ-

5. Economic Foundations

ten in the Koran. 'Allah is the divider of the seed kernel and the fruit kernel. He can pull the living out of the dead and the dead out of the living.' (6.95).

Now I experienced the difficulty I had already frequently met when training the farmers. These people were used to understanding the words from the Koran in an abstract sense and not to think of concrete examples when listening to them. I now showed them using appropriate examples what these verses full of images could mean for practical life. I explained about the millions of micro-organisms and their work in the earth and told them that living earth was connected to the heavens. Then I quoted: 'The sun and the moon pursue their ordered course. The plants and the trees bow down in adoration. He raised the heaven on high and set the balance of all things, that you might not transgress it. Do not disrupt the equilibrium and keep the right measure and do not lose it.' (55.5–9).

Then I asked: 'How can we assist this connection to the heavens? What is the essence of a plant? Is it just a seed we place in the earth, or does this seed receive life from Allah, so that out of it all the different types of plants can grow? Because Allah says: it is not you who cultivates, but Allah cultivates. He lets the plants grow!'

I left short pauses in between my talking to leave space for questions. Then I spoke about biodynamic farming, about the composting process and its preparations and described how the earth was enlivened through them. I explained how we wait for specific star constellations before planting and are thus inspired by Allah to act correctly. Then I led the discussion towards the arrogance of science, which states that it is only physical substances which allow the plants to grow, and not Allah. Because of this people use artificial fertilizers and chemical poisons, never minding the effects they have on people's health and ignoring the consequence of insect infestation.

Suddenly one of the men stood up, came up to me and hugged and kissed me. I noticed how another one had tears in his eyes. What had touched these conventional men? Many were shaken by the concreteness with which one could understand the verses of the Koran. They obviously felt my explanations had deeply acknowledged their religion. After the probing questioning of one of the participants I explained the meaning of the position of the sun and moon for the growth of the plants. And finally, when only three very intellectual participants remained unconvinced, I pointed to the sura 'The Star' (53). This chapter describes how the Prophet received the five praying times from Allah: 'And the eternal part of the human beheld Him come down again, by the sycamore tree, at the end of the path, the repose of the Garden of Eden. And the godly light illuminated the tree. The eyes of the eternal part of the human did not wander, nor did they look into the distance. For he saw some of his Lord's greatest signs.'

What had the Prophet seen? A tree illuminated with light. He saw the living form of the tree, the living processes occurring inside the tree. These processes are reflected physiologically as follows in the plant: when dawn nears, the plant produces glucose until the sun reaches its zenith; this process then ends once the plant's shadow is the same length as the actual plant. This is in the afternoon. The exact time changes depending on the time of year. Between the afternoon and sunset the plant transports the glucose it has previously produced to all its organs. After sunset a third process begins, the plant starts transforming the glucose into active substances. This process ends when darkness sets in and during the night the plant starts to grow. These four processes correspond to the five positions of the sun, and allow the internal life of the plant to appear. The Prophet Mohammed recommended we think about Allah and turn to the supersensory at these five times throughout the day. Because of the relationship of the five prayers with the sun's course and the rhythms of the plant world the praying human connects to the cosmic processes.

I told this all to the assembled, deeply religious people. Then a silence settled over the group. They had understood and experienced that our community consisted of Muslims, who worked responsibly out of the knowledge of the Koran. The whole meeting had lasted three hours. One of the men stood up slowly and said: 'If you tell us that Allah and the Prophet also allow music and art, then we will believe you, because you know better!' I thanked him, asked the musicians to enter again and together we listened to Mozart's serenade, now with united minds and with open hearts.

Before we went for lunch I mentioned how important it is for the farmers to receive the appropriate training, so they can deal responsibly with the earth, the plants and the animals. Then my visitors wanted to know more about our pedagogy and the schooling of the children. The prayer leaders in the mosques had frequently preached against our schools and told parents not to send their children to it. So I took them to a play performed by the children straight after lunch. Some handicapped children also took part. They were deeply

5. Economic Foundations

touched by the joy with which they saw the children move on stage. Afterwards they wanted to know what kind of verses the children learnt at school. I waved some of the children over and asked them to recite what they learnt. Then the men understood that even though the verses were about the sun, we worship Allah and receive strength from him. They apologized for their previous bad words, and started asking other questions: for example, why we kept the children in school for the whole day. This opened up the opportunity for me to talk about the development of a person from childhood to adolescence and the task pedagogy has in this process: I said something like: 'It is important to recognize Allah's signs in all areas of life. And one of the most mysterious signs is the human being. The Prophet describes the different levels of the human being as both heavenly and earthly; but at the same time he describes him as something in the making, thus emphasising the process of development. The Prophet says: the first seven years in the life of the child you should play with it. The second seven years you should learn with it. During his third seven years you should treat him like an adult; not try to bring him up anymore, but treat him like a partner. Then you can let him go in freedom.' We also train our school teachers using these quotes from the *Hadith*. When I continued explaining using more living examples they all felt this wisdom was founded on the depths of Islam.

The grim, bearded men of the start of the day had become my guests said their farewells heartedly and with deep feeling. I knew they would meet again on Friday in the mosques and would spread the word what a mistake they had made. I let them go with the words from the Koran: 'If somebody comes to you and tells you a rumour, then do not believe them, but verify it yourself.' They passed this on exactly. They explained that Islam lives deeply in Sekem, in a way it does not exist anywhere else in the country. And to commemorate their visit they gave us a plaque, written with beautiful calligraphy in golden letters, which states that the community of sheiks verify that Sekem is an Islamic initiative. This plaque now hangs in the entrance area of the school.

After this visit I invited journalists from all the well-known newspapers in Egypt to a press conference. Afterwards they published completely different articles about Sekem. The director of the local paper denounced the disturbing article and sacked the journalist who had written it. The court case against the responsible editors continued and led to a condemnation with compensation for damages. Because I

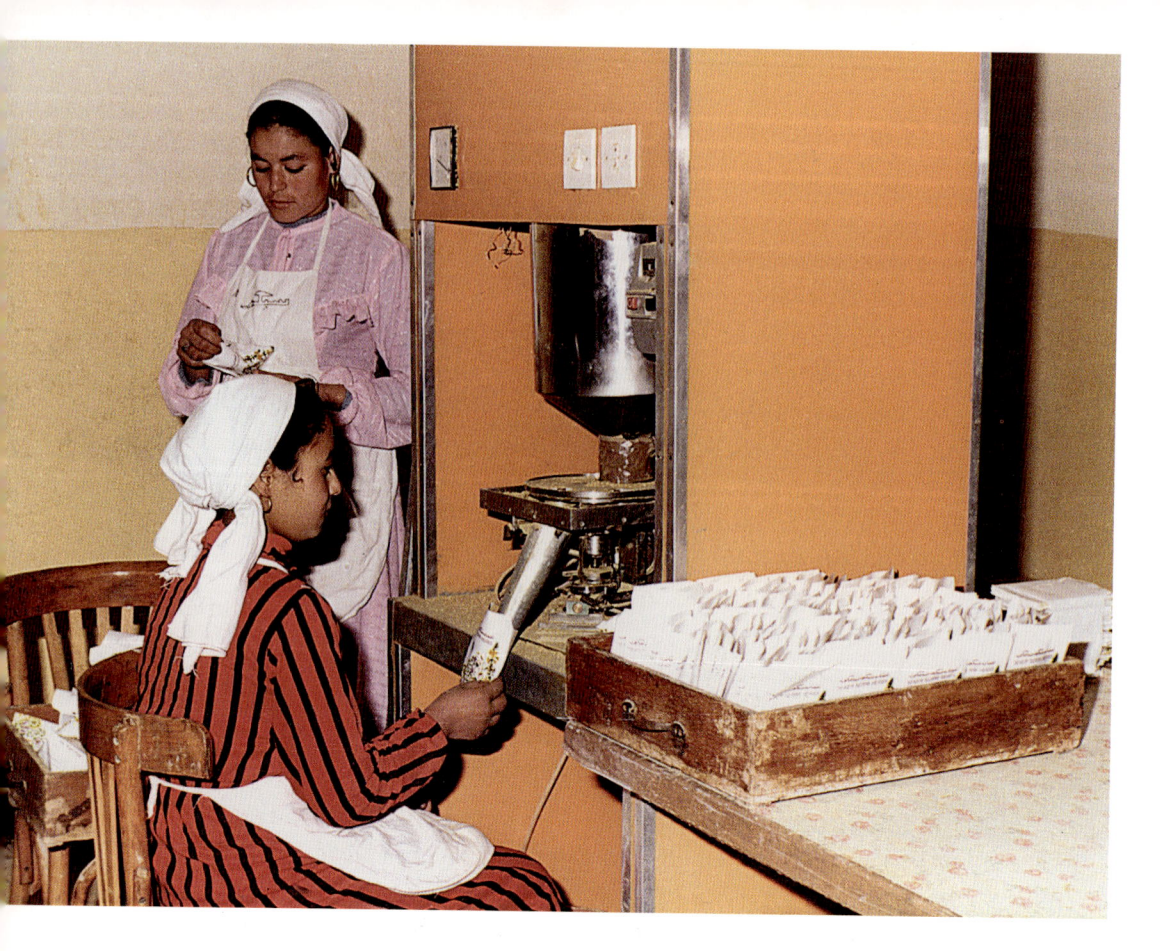

Early beginnings with packing tea, 1993.

forgave them they did not have to pay the damages in accordance with Egyptian law. I was not interested in the money anyway, but rather in rectifying our image.

Pharmaceutical company in the desert

I was incredibly relieved to hold the first licence in my hands which enabled me to run a pharmaceutical company. This secured the economic foundation of the business, and we could establish the rest upon this basis. But it is impossible to imagine the difficult and lengthy process which lay behind me! Nasser's socialistic government had nationalized all private businesses. There were no private medicinal businesses left in Egypt, and when I asked for authorization for my company I was told it was not possible as I was not a

5. Economic Foundations

state business. Parliament had to pass a law that a private business was allowed to produce medicines — and there were huge administration problems preceding this decision! First I had to ask the three largest state pharmaceutical companies in Egypt for permission to run my business alongside theirs. They in turn sent me to a senior university official, who was supposed to assess why the state would need something so useless. One day, shortly before we received the licence, the vice minister of the health ministry asked me to come and demanded an inspection of my business, done by two experts of the largest pharmaceutical company on Egypt. One of the specialists was the business manager, the other his director of research. I was told there were certain regulations for pharmaceutical companies which needed to be adhered to regarding laboratory and production rooms and sanitary fittings and hygiene.

I realized how much the future of Sekem depended on the success of this visit, and prepared for it carefully to be able to meet all the conditions as well as possible. In those days only a dirt road led through the desert to Sekem. If we whirled up too much dust or got stuck in the barren wastes on the way to Sekem my project would probably be doomed from the start. So I made sure the road was levelled by tractors and sprayed with water for the three days and nights before their arrival, then the two men could see how clean the road was kept leading to the pharmaceutical company in the desert. On the day of the inspection I met the men in my office in Cairo with a comfortable, air conditioned vehicle I had organized. I was eager to entertain them during the journey, as they kept asking when we would finally get there — after all, the trip took over an hour. 'You can't be serious that you want to start a pharmaceutical company in the desert? Where will you get the experts and the raw materials?' the manager of the business finally asked. Then I told him about my idea and assured him that everything was well prepared. I also asked them about their products, and thus the time passed until we arrived. First they wanted to see the laboratory. I would not have been able to construct an entire building for the laboratory so quickly, so instead had fitted out a room in the Round House with the simplest equipment. I explained how important it was for the future of research to move to the desert, to be able to start anew without outer constraints. Later I found out that they told others about their experiences with enthusiasm. The attitude of a researcher who thought about the future and went into the desert to implement his ideas had obviously impressed them as much as

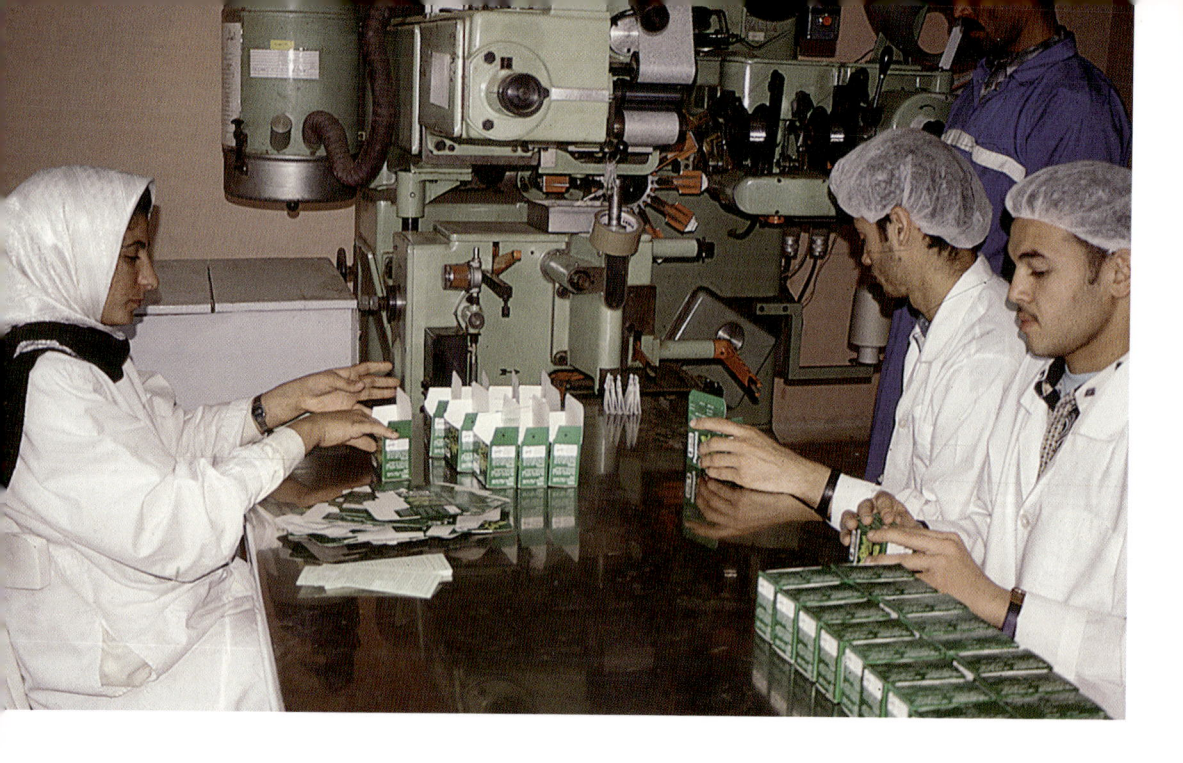

Packing tea for the Atos Company.

the researcher's wife, who bravely worked beside him in the desert. Naturally there were some things they found fault with. But when I agreed to work through their list of requirements they finally gave me the important licence.

Later a salesman said to me: 'Do you realize there are many people interested in starting a private pharmaceutical company? Once you have got your permit you could sell it for a lot of money!' I thanked him for his well-meant advice. But I had other plans. Over the previous decades people in Egypt had forgotten that medicinal plants existed. Sekem reawakened the awareness for these simple but effective health remedies and brought them back on the market. A study into economic viability preceded retailing, conducted by a young management expert, Aiman Shaaban. He carefully researched the market for over half a year and investigated the most prevalent illnesses. Medicinal teas to cure different ailments like kidney and liver problems, coughs, rheumatism and also to support lactation were produced by Sekem. We carefully and hygienically processed the organically grown herbs and filled them into small sacks. When pharmacists asked us for an herbal slimming tea, I blended herbs to make a 'slimming tea' with a diuretic and laxative effect. Extensive advertisement and educational campaigns in the media preceded the first sales. This led to an increasing demand on the market, which in turn means we can still expand our range of products today.

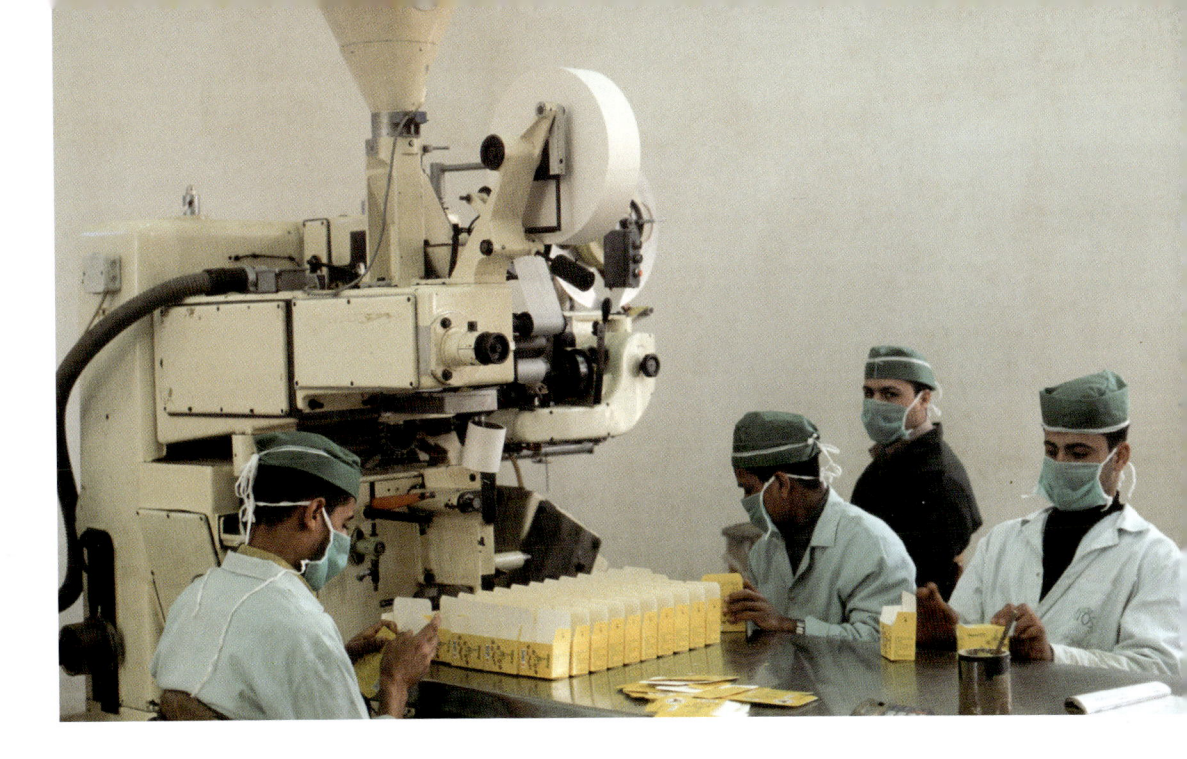

Registering the medicinal teas was another administration saga. Up until then the health ministry had only come across medicines which had a single, clearly defined active substance with a precisely determinable effect. I had to prove that a blend of herbs can also be effective through costly analyses, and persuade the officials that medicinal teas have been used in Europe for a long time. It took a full seven years before the health ministry agreed to register further remedies.

Our first customers were the elderly who could still remember the healing properties of plants. Then the educated people followed, but not, as I had hoped, the poorer classes. We wanted to sell the teas cheaply so as to make them affordable to everyone. Once we had decided on a price and marketed it, the pharmacists told us we were much too cheap. It took years before we found the right balance between the cost of growing the herbs and the market demand. Finally, once the medicinal teas were selling well, we also produced hibiscus, chamomile, peppermint and other herbal teas. I did not market these varieties under the Sekem trade name, but founded the

The Atos Company buildings.

Isis Company to differentiate between them and the medicinal teas. The Isis-teas had to be packed in filter bags, otherwise they would not have sold. So we needed to buy a filter bag filling machine. Christophe and Yvonne Floride and their two small children had just come to Sekem at that time. He and Helmy finally found a very cheap machine from Argentine. It is still working today, along with many others!

Christophe Floride trained as a mechanic in Kassel. He met his wife Yvonne when they were both still at school. His father introduced him to Klaus Fintelmann, and through him he became acquainted with Elfriede Werner. She advised the young family to move to Sekem. First Christophe worked in technology, but during one of my trips to Germany I asked him to supervise the building for the pharmaceutical company Atos. His work was so conscientious that he became our active building supervisor from this point, overseeing the construction of all the buildings in Sekem. Nowadays he manages the export side of the companies Atos and Sekem and looks after foreign contacts. Yvonne is an important figure in the school and is in charge of training new teachers. She is also a talented artist and colourfully paints all the Sekem buildings. I like to call her 'Miss No Problem' because she performs all her tasks selflessly and with ease. I cannot imagine Sekem without them and their two children.

Many young people also come to work in Sekem, either as a part of their practical training, or to absolve their alternative to military service. They have the chance to experience this usually completely foreign culture for some time and to work in the economic and cultural organizations. Christina Boecker came to us after her school leaving exams to get to know the economic side of the initiative. After she completed her business management studies in Germany she worked for many years in our export. Now she lives in Germany and manages the entire European export from there. She also regularly edits the *Sekem Insight* newsletter, which is also available on the Internet. A growing number of interested people can now keep themselves informed and up to date about developments in Sekem. Tobias Bandal came to us for his community service, planning to study physics afterwards, until he realized that the world can be built up using economics. He studied agriculture in combination with business management while working together with Sekem, and is now preparing for future tasks.

Atos: the phyto-pharmaceutical company

I had never given up the search for a financial partner for Sekem since the agreement with the Bank of Islam had failed. Then one day I discovered that the German Developmental Company (DEG) was trying to attract developmental projects in Egypt. One of the conditions for application was a German partner having a share in the company. I remembered my encounter with Roland Schaette, the manager of a veterinary medical company whom I had met in Bad Waldsee after my Italian journey with Georg Merckens many years ago, and for whom I had felt an immediate liking. Roland Schaette was willing to take part in a joint venture company. Together we founded the pharmaceutical company Atos Ltd with the DEG.

The first remedy which Atos developed out of this cooperation was called 'Tomex.' It was a highly concentrated garlic powder preparation and we bought used presses and machines from Germany to make the pills. At first all the workers couldn't stand the smell of garlic. Nowadays Atos Ltd fulfils all the hygienic regulations of European pharmaceutical companies.

To further establish the company we hired experts from state institutions, who helped us with their expertise and built up a capable production team. For many nights I sat at the computer with my friend the architect and designer Winfried Reindl to create the packaging image and to devise the advertising campaign. The basis of all new preparations of Atos Ltd was prepared through extensive marketing studies.

One day the health minister demanded clinical trials to prove the effects of phyto-pharmaceuticals, as Egypt had lost its awareness of the healing properties of plant remedies. We had to have tests conducted by the universities. We gave lectures to stress the importance of herbal remedies for the country and to educate the professors about the side effects of allopathic medicines. Some of them were convinced by our talks and supported the struggle for introducing herbal medicines. But we had to win over each one singly.

The director of the registration office, Professor Haider Ghalib, was a pharmacologist as well as a paediatrician. He publicly opposed the use of phyto-pharmaceuticals, because according to his view there was a danger of fungi on the herbs. He declared the secondary metabolites of the fungus, the mycotoxins, were toxic for the liver, and because of this one should never give children aniseed or caraway seeds so as not to damage their liver. I frequently tried to invite him to Sekem, but he refused. Eventually we did agree on a date. When he visited us his rejection was written clearly all over his face. First I showed him the pharmaceutical warehouse and explained the procedure that we always use, alternatively gassing the herbs in chambers with carbon dioxide and oxygen. I reminded him that nothing living could survive the carbon dioxide and thus all the micro-organisms and fungi die. The following oxygen treatment caused any remaining germs to grow again, which were then killed off with a further treatment of carbon dioxide. This method, developed in Berlin, clears the medicinal herbs of any damaging micro-organisms. The exacting scientist observed everything closely and finally looked at the apparatus with great enthusiasm, hugged me and exclaimed repeatedly: 'I can't believe something so brilliant exists!' Then I showed him around the company, pointed out our hygienic procedures and also explained which ethical principles we followed. When he found out how we trained our workers his joy knew no bounds. Arm in arm we went around the farm, and I told him about my vision. He could not resist the opportunity to bend down and feel and smell the compost. As a doctor and a philosopher

5. Economic Foundations

he could immediately understand and appreciate our holistic initiative. He became one of our most enthusiastic defenders in the ministry of health and is still actively and amicably connected to Sekem.

Later I asked Professor Ghalib, as a member of the ethical medical committee, to investigate the use of mistletoe, a tried and tested medicine for treating cancer in Europe. We wanted to cooperate with a European company which had experience with mistletoe research and producing mistletoe preparations. We discussed which renowned professors would need to work with us, and presented our interest to the former health minister Dr Mahmoud Mahfouz, an old acquaintance of mine. It was necessary to overcome the great prejudice of the doctors and scientists who could not imagine how a plant could have any effect on cancer. I arranged for Professor Mahfouz and Professor Ghalib to visit the Carl Gustav Carus institution and the clinic in Öschelbronn which specializes in mistletoe treatments in Germany. There they met up with Dr Armin Scheffler of the Carus institution and Dr Hans Werner in Öschelbronn for discussions. They came back full of enthusiasm and willing to attempt to register mistletoe

Work in Sekem Medical Centre laboratory.

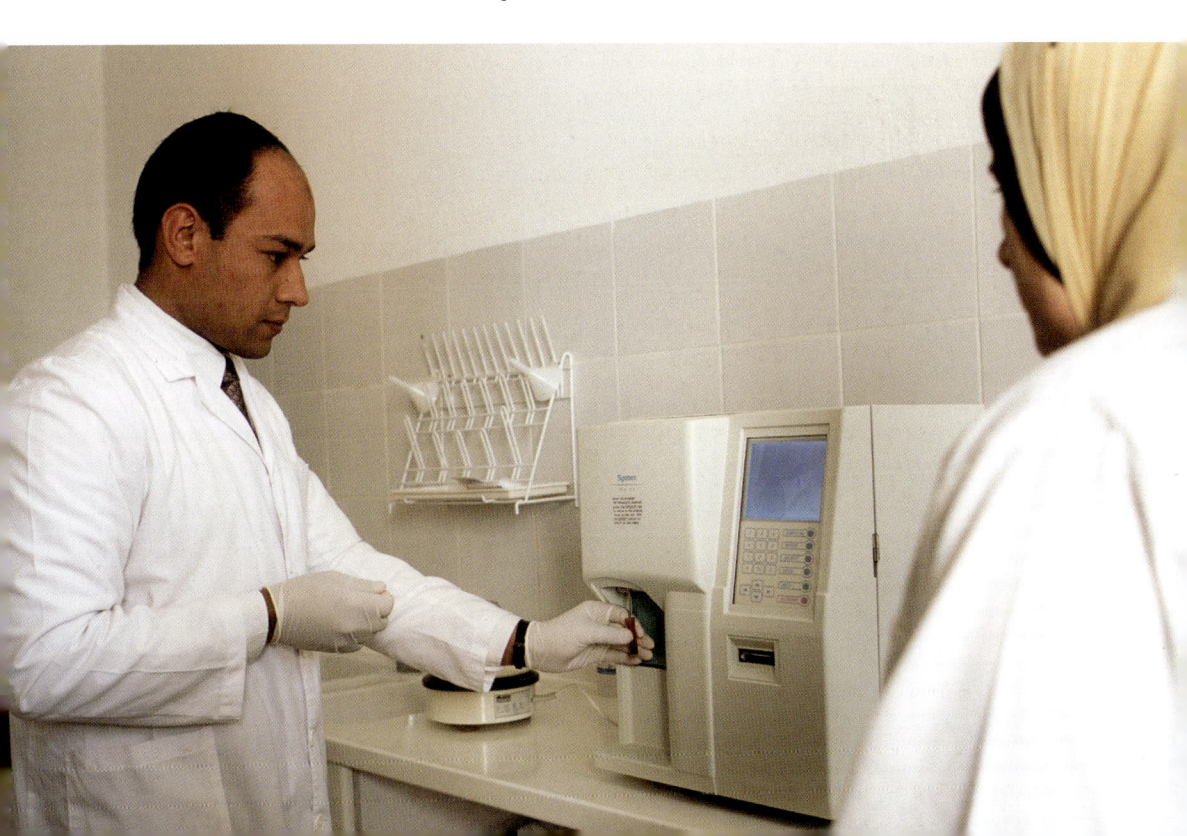

medicine as a healing remedy in Egypt. The work on a report for a preparatory clinical study took six months. Finally we could present it to the ethics committee. We invited twenty oncologists from universities throughout Egypt and organized a meeting in Cairo. Armin Scheffler and Hans Werner also took part. We managed to convince almost all the doctors to take part in the trials, which started soon after in oncology centres in Alexandria, Tanta, Cairo, and in Upper Egypt. The results of the trials proved to be so unexpectedly good that the officials agreed to register 'Viscum,' as the remedy was called. The Atos company has been making a Viscum preparation in two concentrated forms for several years, one of which is used for treating Hepatitis C. The marketing of Viscum launched another discussion about prices as all medicine pricing is still in the hands of the State. Further studies with mistletoe followed, requiring a lot of effort and expense. At the moment Sekem employs ten doctors to advise the professors using the preparations in the oncology centres.

Nowadays herbal remedies are in great demand in Egypt, and we are looking for experienced managers for the further expansion of Atos. If the market wants to buy our products, we have to invest in them further, which requires capital and capable people. My vision always encompassed covering the basic needs of the people by producing food, medicine and clothes, and with these products working in a healing way.

The medical centre

Some years after founding the school the efforts invested in the medical area had matured to the point they were ready to take on a concrete form. I planned the rooms for an out-patient medical centre with Dr Hans Werner. We decided the building should be constructed with an opening gesture and an artistically formed inside, so that patients entering it attain confidence and feel surrounded by beauty. A playground in the garden of the medical centre shortens children's waiting time. Sekem received one of its most beautiful buildings with the medical centre. The inner enclosure has been planted with aromatic medicinal herbs and is used as a waiting room for the patients. Medical treatment is offered in different consultation rooms: internal medicine and paediatric care, ENT, laboratory, operating theatre and X-ray as well as a pharmacy. The treatment rooms are fitted with medical equipment

5. Economic Foundations

from Germany. Dealing with the formalities necessary to import the equipment was again incredibly awkward and led to a struggle with the Egyptian customs. The partially older models were thoroughly overhauled after arriving in Sekem. We finished the medical centre just in time to hand over the key to Hans Werner on his seventieth birthday. All the farm employees gathered in front of the new building for this event to congratulate him. The doctors came from Cairo and other larger cities of the country, and Hans Werner took over their further training. Despite his advanced years he still learnt English for this task and has become a great example for the Egyptian doctors. Together we made contact to the surrounding hospitals, in case we needed their help with further treatments or operations. Many more than one hundred patients use the medical centre daily, which desperately needs to be expanded.

It was as difficult to obtain official permission to build a medical centre in the desert as it had been previously to establish the pharmaceutical company and the school. A group of people had to become informed about the necessary administration and legal requirements, discuss the plans with the bureaucracy and then negotiate with the health ministry. We almost failed to establish the pharmacy, as the bureaucrats insisted we needed fresh water plumbing, but we only had well water in the desert. This forced us to let our water be inspected once a week, but finally we got the necessary authorization.

We received a generous present from someone in Germany for the inauguration of the medical centre — an ambulance! We employed three social workers to drive to the villages once or twice a week with doctors to take medical stock. The houses were numbered, the names, age, income and status of the residents taken and the doctors performed examinations — we wanted to have an overview of the health of the people in the area surrounding Sekem. We found out that 86 percent of the population had parasitic illnesses, among other usually chronic ailments. It became clear to us that on top of the medical treatment centre we also needed to educate the population about hygienic practices. Basic knowledge about hygiene seemed to be completely lacking. The social workers started visiting the village elders and announcing lectures through the mosques, where they discussed themes like hygiene whilst preparing food and controlling vermin. We also noticed that many houses did not have any sanitary facilities. With a 'WC project' we helped the population to dig trenches and

The medical centrer with the donated ambulance.

install toilets and sanitary fittings. Every year we had inspections for parasites, and gradually we could record an improvement.

Sekem also supported a project involved with births in the area, which often occurred under terrible circumstances. Many women died after giving birth because of the miserable hygienic standards, and the babies died of infections or diarrhoea. I asked Dr Roland Frank, an obstetrician from Vienna, to come to Sekem and invited all the midwives of the surrounding area. I also sat at his lectures, as I had noticed a long time ago that people felt motivated by my presence. I was only too happy to remain if it meant they were more open for new ideas. I introduced these lectures by talking about Allah, who created the world and within it the realms of the minerals, plants, animals and humans in such a way that these realms could communicate with each other and together create a great unity. Because of this plant remedies are surely more beneficial to healing than man-made synthetic substances with side-effects, because people do not have the creative powers of Allah. They cannot create and keep the balance found in nature. Allah says: 'We will show them our signs in the cosmos, in the earth and in themselves; using them they will recognize that He is the truth.' And further: 'Healing comes from Allah and not from the substance. He who has created me leads me along a straight path; He nourishes me and gives me to drink and when I am ill He heals me.' Allah demands we find healing remedies in nature. We need to learn to read and recognize his signs.

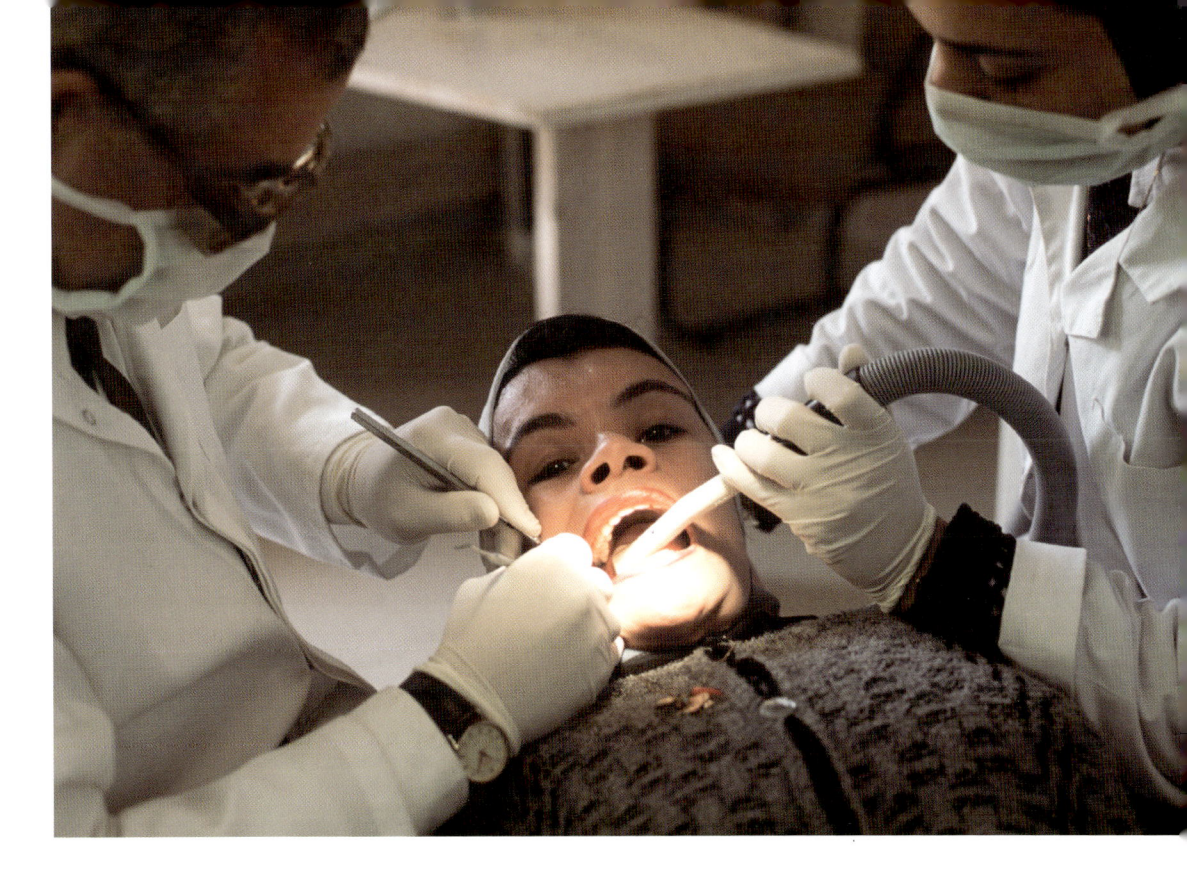

    I realize that this education work will take decades, if not even longer, before results are noticeable, as the people have become too accustomed to living beyond what is healthy. They have little knowledge about hygiene, food preparation, children's education, looking after the senses, land and vegetable cultivation, and clothes corresponding to a natural need for warmth. Although all Sekem employees are educated in these matters, their habits at home are so strong that any change is very difficult. So we see our task as educating on both fronts. The social worker team in charge of this project is very motivated. About thirty thousand people live in the area around Sekem, including several thousand children, who naturally do not all attend our school. We have made it our task to examine and treat these children regularly. The medical centre bus carries one external school class a day to the out-patient centre. Because of this there are lots of children in the medical centre each morning; playing in a prepared area until they are examined.

Ingeborg Marienfeld is a great support to Hans Werner in the medical centre. She is a nurse who moved to Sekem from Germany seventeen years ago with her husband Dieter Marienfeld. All people arriving with a greater or lesser ailment go to see her first. She helped establish the Mahad, taught the girls to sew soft toys and dolls and is now devoted to the training of the employees in the medical centre. Her husband is responsible for any problems concerning the electrical or technical supply of the houses and buildings. Dieter is the technical director of all the janitors in Sekem. He has never lost the courage to meet people with great charm, despite all the difficulties he has encountered with the Egyptian workers.

6. Education and Culture

'Sekem is one large school'

We had to continue building while developing the economic sector. Sekem has had its own building team consisting of about eighty people since the start of the venture. I was interested in the particular building materials used in Egypt, and visited a factory which produced pre-stressed metal. Two engineers, Abdel Hedi and his partner Sami, developed a machine to pull metal into a grid. I had long fascinating talks with these interesting people on several occasions and also invited them to Sekem. They started to get inspired by my vision, although initially all they could see was desert. Abdul Hedi's wife worked in nuclear research. She and some friends asked me to tell them more about the background of my work. Soon we were meeting up nearly weekly in a hall in Cairo. The event always started with a lecture I gave, followed by discussions and questions. Music was played at the start of the evening before my lectures, and this is still the case today. I talked about pedagogy, medicine and agriculture, but also about different themes from anthroposophy. These meetings led to the establishment of an association, which sponsored all the cultural institutions planned as part of Sekem. Although we spoke Arabic in the meetings, Elfriede and Hans Werner often took part and also brought their friends from Germany, who enriched the meetings with their contributions. This group of artists, scientists, politicians and pedagogues grew together over the course of several years and founded the Egyptian Society for Cultural Development, SCD, in 1984.

The trees on our grounds had gradually grown to a good height, the first buildings of the pharmaceutical company were erected, people came and went on Sekem and the cultural and social sectors were developing. One day Angela received a visit from her father Georg Hofmann, who was a teacher in the Waldorf School in Stuttgart. He observed the individual support given to the people working in the companies and agriculture, how they were treated and trained, and listened to our plans for the future. Then one day, when he was standing in front of the Round House, he suddenly exclaimed with his rich, deep voice: 'Sekem is one large school!' What did he mean by saying that? What had he observed concretely?

One day hunchbacked Ali, who was chased out of his village, turned up at our place. I gave him simple garden work to do. With his arrival we started working with handicapped people in Sekem. Later the deaf and mute Hassan followed and Zacharias, who has a mental handicap. Fathy, tall Ali and Scharafa were in their twenties by the time they came to us, some of them married with children. I always welcomed them as if they were friends. Initially the engineers delegating the daily work refused to employ the handicapped people. So I trained them myself and practised their duties with them over a period

Annemarie Ehrlich: Eurythmy in the work place.

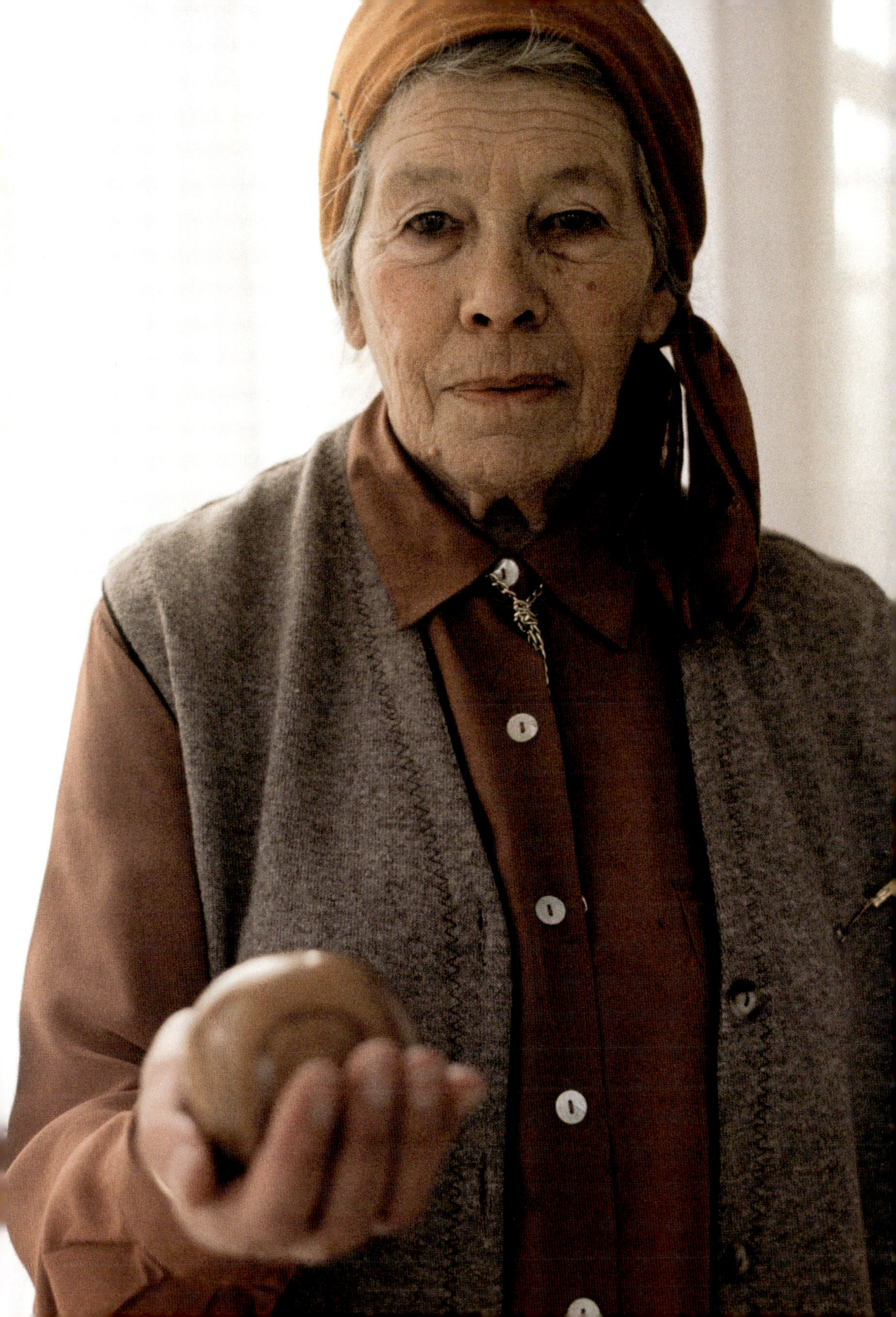

Employees participating in geometric drawing training, as part of adult education.

Sekem kindergarten.

6. Education and Culture

of time: I showed one of them how to irrigate a specific area, another one I helped to muck out the stables, the blind one levelled the paths, strengthening them with sand and gravel and raking them smooth. Hassan helped with the clay building. Everyone received their task, and because we treated these people normally, the Egyptian workers also got used to them over time. Sometimes we were called the 'farm of the handicapped,' which I took as a positive statement of how the surroundings saw us and what we radiated. Later we started our own special education school for handicapped children.

Because agriculture is very labour-intensive we could always employ many people, which at the same time gave us the opportunity to train and educate them. Without support these people would have remained unemployed and, as unfortunately happens often, fallen prey to fanatical groups because they felt the meaninglessness of their lives. We employed waifs and strays to pick cotton or flowers. Some of them had never been to school, or had been forced to leave as their parents needed their help to earn money. When such uneducated young people grow up without support they become ignorant and uncouth. We gave them something useful to do and also made it possible for them to go to school. There they not only learn to read and write, but also to sing, paint and do eurythmy. They are also taught information about hygiene and healthy nourishment. At the same time they receive full pay for their work.

We founded the project 'Chamomile Children' in response to the terrible and widespread social situation of child labour in Egypt. The Chamomile children, aged between ten and fourteen years, have their own teachers who look after them the whole day. As well as schooling, they work picking herbs and receive a cooked meal every day. Medical care, which is particularly important for the often impaired condition of their teeth and eyes, is also given to them. After attending the school they receive a report, like the other children, confirming they are not illiterate anymore. They then have the chance to learn an occupation in the vocational school. This gives them a better start in life and they can devote their time for the good of society. To reject all goods produced using child labour out of principle is not the lasting solution to this problem.

In 1986 we started planning and building the Mahad, our centre for adult education, to give an institutional framework for our intent to educate adults. The concept for the white round house was created by Winfried Reindl: a building sheltered by tall, dark casuarina trees,

Gudrun Abouleish with handwork students.

with a meeting room and two teaching rooms. It was festively inaugurated in May 1987. Here we also educate children with handicaps. The girls working in the companies learn to eat nourishingly and dress appropriately and hygienically through courses given in the Mahad. Knitting and embroidery helps to reinforce the sense of beauty of the fellahin girls. These procedures have become known throughout the surrounding area, with the result that the girls trained in the Mahad are very sought-after wives.

We built a kindergarten large enough to encompass two groups beside the Mahad in the same architectural style. This started Sekem's actual pedagogical work in 1988. The first Egyptian kindergarten teachers were trained by Konstanze.

Since 1985 I had been travelling around European cities several times a year with Elfriede Werner to talk about the Sekem initiative. One day I was giving a talk in Munich in front of many people. After the lecture some people came up to me to ask questions, among them Regina Hanel, who asked if she could come to Sekem. She started working as a kindergarten teacher, until it turned out several years later that she also

Above and overleaf: School life.

had a talent for typing, organization and administration. So I asked her to take on secretarial duties for me in the administration office in Cairo for three days a week. She has been doing it since with admirable reliability. She is an enormous help to me with her ability to create order and a protective space for me to work in. She still supports the Egyptian kindergarten teachers for her other three working days.

We started our school in 1989 with a class one and a class seven — a class each from the lower and the middle school simultaneously. Twenty-seven Class One pupils walked hand in hand in a line, their heads decorated with colourful wreaths. They were led into the Mahad by their teacher with the Class Seven pupils following behind them. Their parents had already gathered there, fellahin and Bedouin, discover what awaited their children after the musical introduction. It is very important to us that the parents support their children. Regular parents' evenings teach them the importance of rhythm for children, for example, and also that the widespread, uncontrolled watching of television is unnecessary for children. They are also taught how to make sure the children have good nutrition, clothing, and much more.

Kindergarten chiildren. Below with Regina Hanel.

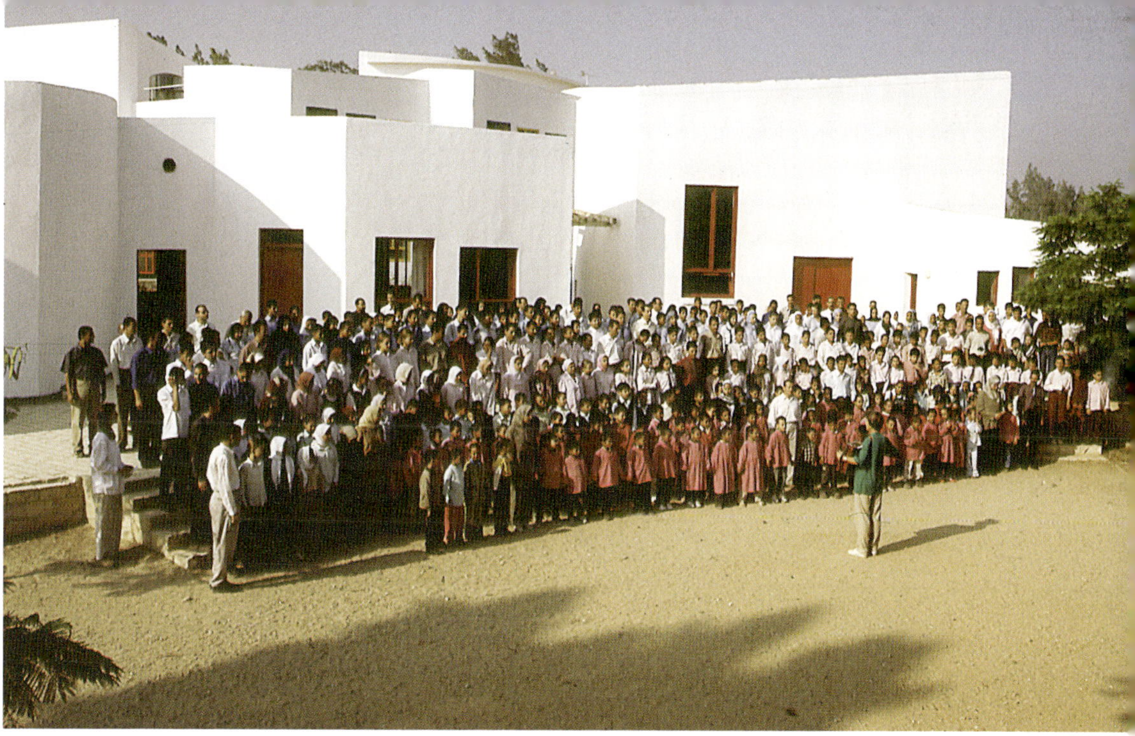

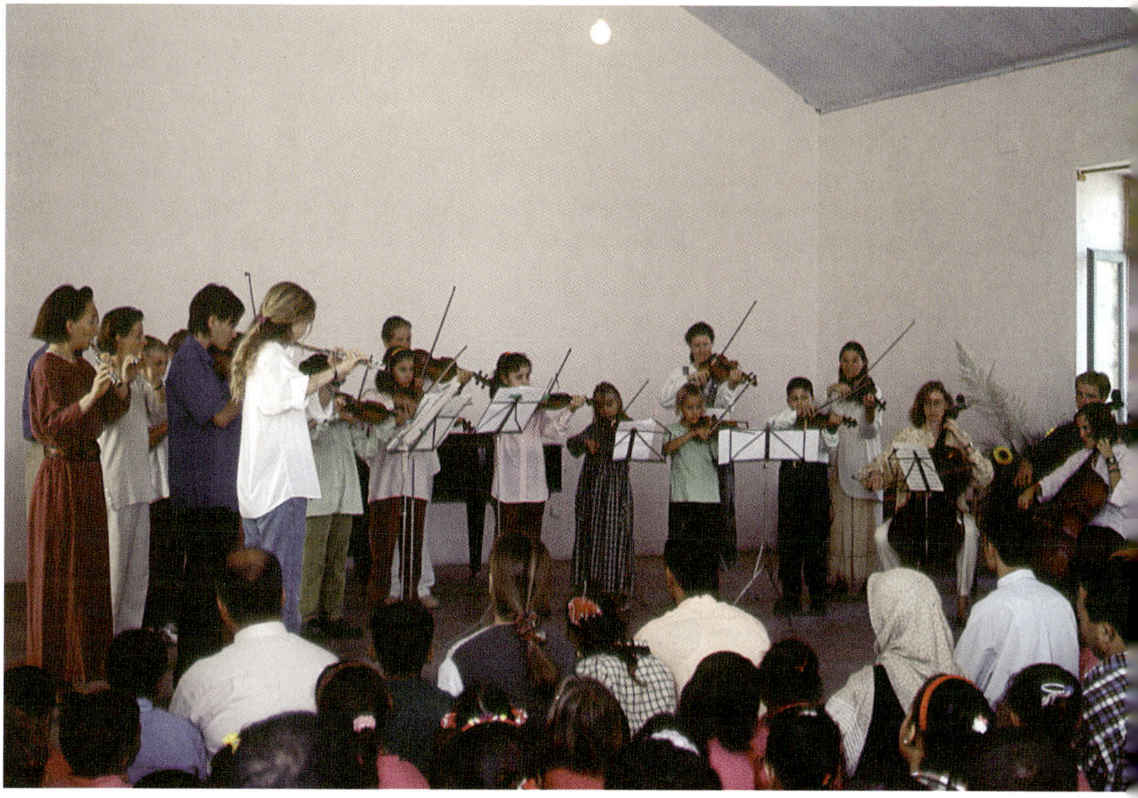

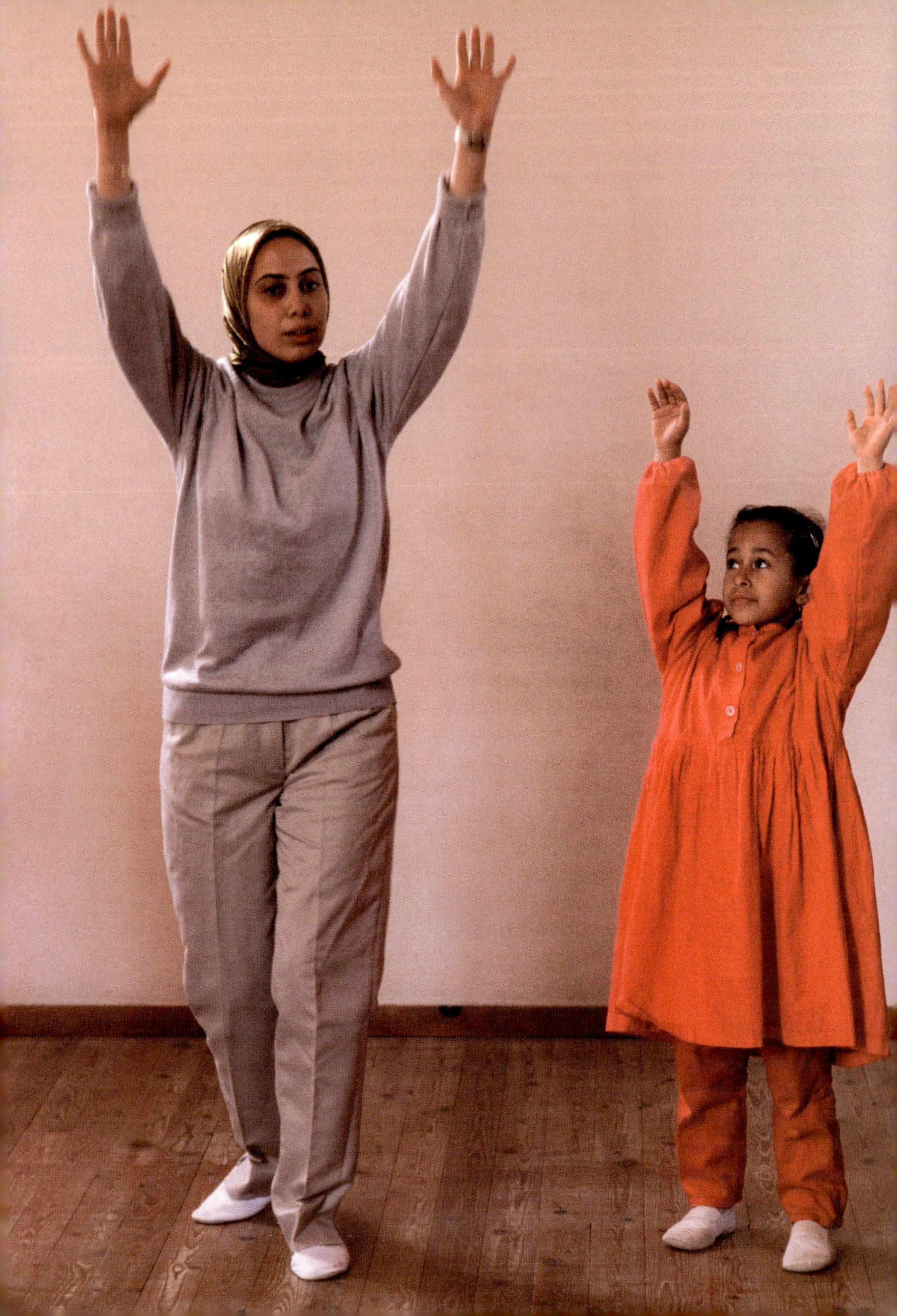

6. Education and Culture

During this time I spent alternate days in the administration office in Cairo and on the farm, dealing with the thousands of questions concerning the foundation of the school. I used each spare moment to establish the school, as well as sorting out the financial questions. The financing was set up so that the school was sponsored by the companies, that is, it was part of the fixed investments of the companies, which they rented to the Egyptian Society for Cultural Development. I borrowed money from banks to finish erecting the school buildings. We also received financial support from our German Association for Cultural Development in Egypt. Elfriede Werner worked untiringly to organize lecturers who supported us by giving training courses. She collected money for all the necessary furnishings for the school and for lesson materials. I have already mentioned Professor Klaus Fintelmann, the pioneer in the area of training apprentices. The Hibernia School in Wanne-Eickel, which he co-founded, is a comprehensive school in which vocational training is integrated into the curriculum of the upper school. I was able to see its advantages with my own eyes. Together with Klaus Fintelmann I penetrated deeply into questions of pedagogy during the following years and we devised possible pedagogical concepts for Egyptian conditions. We also asked Winfried Reindl's advice, and made clay models for a school building on a model landscape. During the day we rejected the ideas of the previous night, as new aspects became more important. It was a very lively time. I remember we decided that the kindergarten and first three classes should be in the same building so that children of a similar age group remained together. We planned the shapes of the different class and subject rooms and their colour schemes, teachers' rooms, parents' rooms, and a doctor's room, as well as a chapel for the Coptic children and a mosque.

But it was only in 1990 that we managed to find ten days to implement the developmental project. During our very lively discussions about human foundations and their transformation towards shaping concrete forms, a young architect, Gerdi Bentele, became involved and quickly converted our ideas into sketches. We interrupted the discussions to look at the sketches and joked about our ideas. I remember we thought about integrating playing areas into the building of the first three classes — sandpits and water features — all things unusual in Egypt. Gerdi quickly sketched two trees on to the paper as well, and we all starting laughing full-heartedly, as we could see the children climbing all over the trees in our mind's eye. When the sketches were finished Gerdi brought us clay and we started making three-dimensional forms

Sekem school.

following the sketches, taking the slightly hilly landscape into account. After we had finally created a vivid, three dimensional model out of clay, I had such a strong urge to start building that early the next morning I began levelling the planned area with huge bulldozers — although in retrospect I can see that for an architect the actual work only starts at this point, even if the idea sufficed for me. When Gerdi, Winfried and Klaus got up the next morning, they were extremely surprised to find tractors removing earth and sand to dig the foundations for the school!

The school building was built in three stages with a painting and music room, several handwork rooms and eurythmy rooms as well as the twelve class rooms. The school also has a large hall, used for end of week assemblies, where pupils perform from their lessons, celebrating the festivals, and holding employees' question hour. The area surrounding the school was levelled to build a huge playground and gymnastic grounds.

We invited the officials of the education ministry to Sekem for a day as part of the registration process. It had not been easy to convince them of the necessity of a private school in the desert, as up until then there had been few private institutions in Egypt.

The meeting began with a recital from the Koran, followed by classical music played by our musicians, which was just about tolerated. Then I explained each musical instrument and the necessity of good music lessons for the education of the children. I also spoke about the importance of movement elements in the lessons, and the meaning of artistic work: 'The children need a balance between theoretical learning and artistic and practical work to satisfy their real educational needs. Training their thinking and learning by heart are not as important as imparting real experiences through many different education elements.' The officials were able to understand what I said when I reminded them that school education used to be a lot richer. 'Yes, that is what it used to be like in Egypt!' They agreed. 'Unfortunately it is not like that any more! If this new school really works according to these principles, then it will become an example for our whole country!'

School building, vocational training centre and mosque.

Then one of them stood up and asked: 'And what part does the Koran play in your education?'

Learning the Koran by heart is an important aspect of Egyptian schooling. 'We want to impart reverence for creation in all the lessons,' I answered. 'The feelings aroused by learning are most important.'

He was not content with my explanation. 'Yes, but the Koran?' and his voice cracked: 'What about the Koran?'

'We will also teach the children the Koran, but not excessively. They will get to know the prayers and images, but learning by rote in a one-sided way only caters to the intellectual abilities of a person. We strive towards instilling a reverent attitude towards all creation, not just a specific religion, and thus develop religious feelings in the children entrusted to us. That's a lot more valuable!'

I never acted provocatively when talking to people about such delicate matters, and also never contradicted them, because their thoughts and behaviour were basically consistent with their upbringing and with those prevalent in this country. I tried to impart understanding, and start the discussion from their point of view. My task consisted of broadening their views and trying to instil in them the idea that there was more to the issue than they were used to thinking. This

Klaus Fintelmann and Ibrahim Abouleish.

respectful treatment created a feeling of trust, and out of this trust we finally received permission to found a private school. The problem of a school director with the necessary qualifications was resolved by my wife Gudrun, who had attended the teacher training course in Graz; she was accepted by the government for this position. The school underwent frequent inspections, and she always endeavoured to solve problems and find solutions acceptable to everyone. The criticisms of the officials were also partially helpful, for example when they asked for detailed lesson plans, which the teachers had to supply.

The next problem was where to get the many teachers necessary for the new school. Sometimes we had to use unconventional methods. Mohammed, for example, had reliably administered a warehouse for many years. I valued him as an open, polite, patient man. Before coming to us he had finished his teacher training course, but never actually taught before. One day I asked him to come to me and told him: 'Mohammed, you are going to become a teacher at the new school!' He was not as appalled as one might imagine — he could have rightly objected that he did not know anything about pedagogy and had never stood in front of a class, even though he had studied teaching. Instead he felt honoured and recognized by my request and completed all the

Sekem school.

training requirements with interest and joy. Training our teachers was very time-consuming. I worked with the future teachers for several hours three times a week, covering questions about human nature and the Koran. The intellectual abilities of a teacher were of less importance to me than their character, which works on the children in a humanizing way. Our training course was enhanced by guest lecturers from Germany. I am very thankful to them for helping to establish our school. Giving lectures to the teachers is still one of my tasks at Sekem.

Celebrating the laying of foundation stones, topping out festivals and inaugurations with the community comprised of children, parents and friends were wonderful and important milestones on the path towards implementing the vision of Sekem. I am happiest when I see that our efforts are good for the children, that everything is not just a beautiful idea, but means concrete developmental support. The weekly highlight in this context is the end of week assembly on Thursday afternoons. These celebrations are the result of intense work and the concentrated, united atmosphere of so many hundreds of people is almost a miracle in Egypt. During the many hours of training the teachers I had taken pains to point out the importance of keeping the children quiet during the lessons and when gathered together in a group, so that the lesson content and atmosphere can be imparted suitably. We practised this with the teachers and they then passed it on to the children, so that now, when I stand on the stage and greet the children with *'Salem aleikum,'* their answer comes out of a rich silence.

Then the Koran singer sings his song with beautiful voice into this silence, which develops the religiousness of the children and speaks to their national traditions. The musical performance which follows trains the children to listen attentively week after week. We carefully attend to the quality of the children's contributions from their lessons. I used to always watch everything at the dress rehearsal, and make comments and corrections. I am always moved to see how confidently, joyfully and independently the children stand on stage and practise articulating themselves clearly and without constraint. When the end of week assembly is finished, all the children pass by me and shake my hand. I say goodbye to them and look at their appearance, feel their hands and observe their gaze. If I notice something amiss I can then discuss it with their teacher.

Sometimes I use these gatherings to give a short speech about themes involving the whole Sekem community. For example: we built a small chapel for the Coptic children in the school, where a priest celebrates mass with them. I once told the assembled community that the Coptic children needed a cross for this chapel. I tried to explain this foreign, but beautiful form to the Muslim children by showing them that every human is able to make the form of the cross with their limbs, and how the earth itself also has a cross on it in the form of mountain ranges. Then some Muslim children eagerly raised their hands and asked whether they could carve a cross for the chapel.

At a different assembly I talked about the mosque on our grounds and told about its beauty, which requires cleanness. The Coptic children decided to clean the mosque on Fridays, while the Muslim children still decorate the chapel with flowers every Sunday.

The Vocational Training Centre

The lack of trained experts is one of the main obstacles for Egypt's further development. Central Europe's deeply rooted craftsmanship with its apprenticeships is not systematically employed in Egypt. I also realized how important it was for the young people in the villages to receive work and have training. In 1997 we were able to found a vocational training centre. At the end of the obligatory school years the adolescents can train in one of the following professions for three years: metalwork, carpentry, mechanical work, electrical work, tailor, biodynamic farming and trading. This development started us off in a new direction! After initial searching we suddenly found unexpected possibilities opening up. A project founded between the Egyptian President Mubarak and the German Chancellor Kohl, named MuKo after the initial letters of both ruling leaders, helped us with the foundation. We were given the required curriculum for the vocational training from the German Society for Technical Cooperation. Because the necessary machinery came from Germany, we were helped with all the official and customs formalities. The state also received public money for inspecting and certifying the school.

Because of this help we were able to open the Vocational Training Centre fairly quickly. Now it has developed into a school with 150 pupils. Each year has twelve trainees in the different vocations. Initially the small number of pupils made it the most expensive venture which our sponsor society (SCD) ever had to finance. Even though the companies covered the training costs, money for clothes, food and some pocket money during the training had to come from the SCD.

Conytex printing department.

It is a miracle that the cycle of money always works out, since helping people is always the most worthwhile investment.

The school was established following the so-called 'sandwich system,' where pupils receive theoretical lessons and learn on model machines, but also work in fully equipped workshops where they receive the necessary practical training. I would like to mention Klaus Charisius, Wilfried Ulrich and Eberhard Kläger in this context and thank them for their untiring work. They visited us frequently and supported us with their years of experience. Through the social inclusion the pupils gradually matured into adults. After finishing their three-year training they usually know exactly where they want to go next. Many adolescents are employed in the Sekem companies as carpenters or as mechanics, if they do not have to do military service. Others return to Sekem after their military service or become self-employed. Our graduates are valued workers and it is always a joy to observe their further development.

Vocational training.

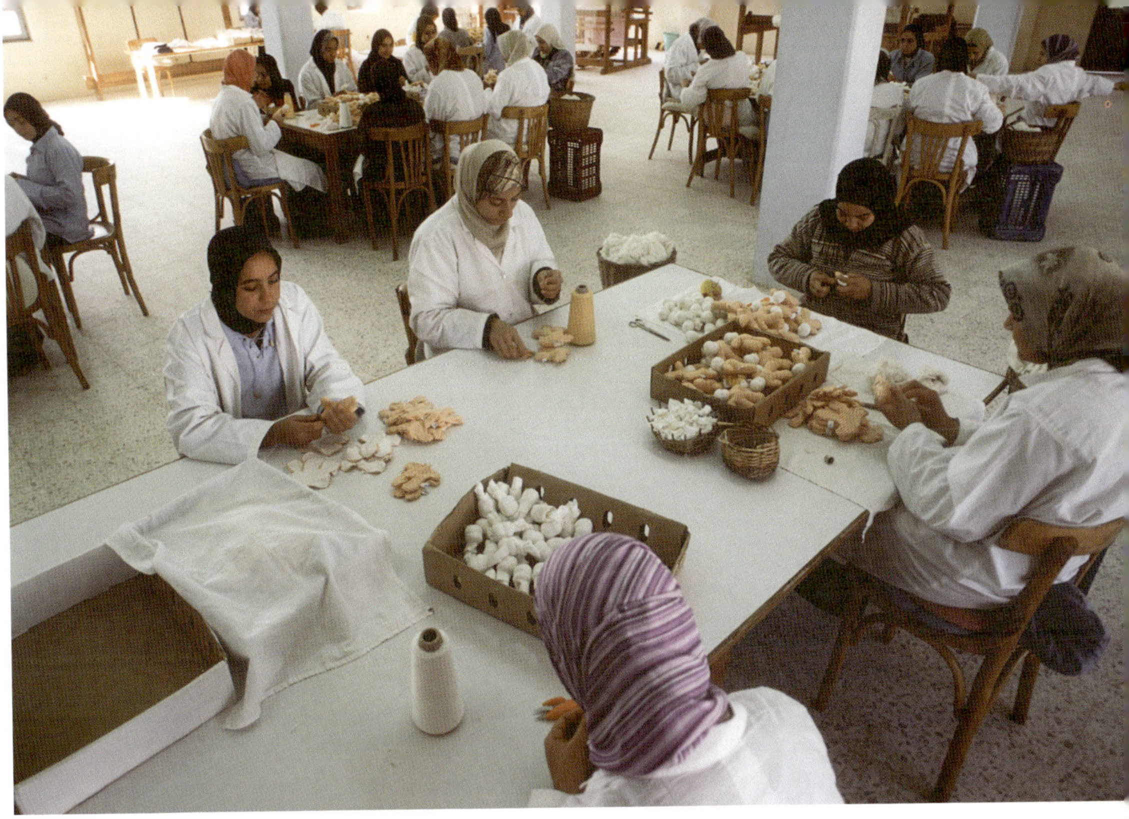

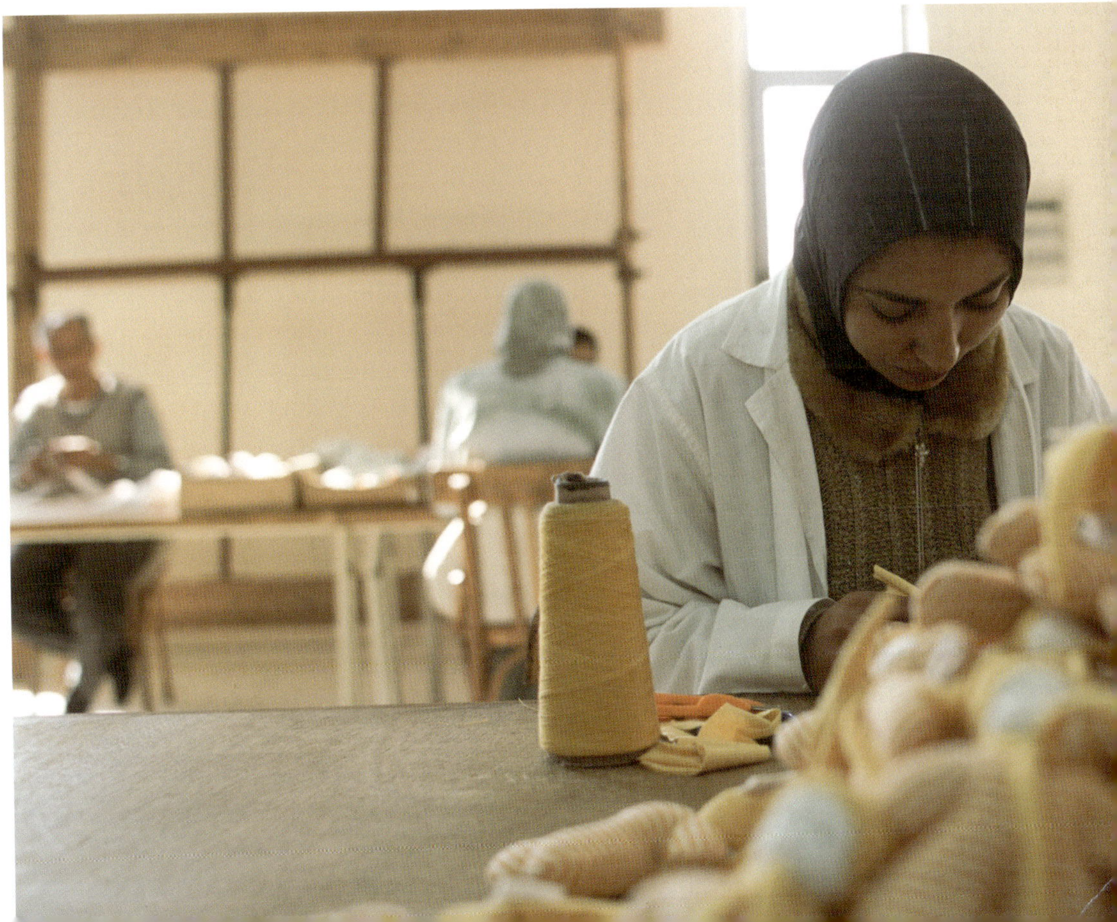

The university

I conceived the idea for the Sekem Academy, as I called it in those days, on the same day as the vocational school opened. But in 1997 there were so many other projects in a state of development needing my support that it was not possible to set up anything new. Added to that I felt the idea for the academy needed time to mature. The time was not right yet and I did not want to be too hasty — but I still wanted to act. So I started by planning the rooms, as every idea needs an institutional framework for it to be implemented. The first building for the planned academy was finally inaugurated in 1999 in Heliopolis. This building — seen as a part of the whole planned Sekem University — was intended as a library. Applied research in different faculties is conducted in the rooms, and there is also a large hall that is used for administration meetings with all the staff. Sekem has always conducted research. But over the years the research has become increasingly concrete and specialized into different departments. Since 1999 the Sekem academy offers applied agricultural, pharmaceutical and medicinal research. It also conducts studies and events in the areas of economics and law. I know that the planning phase for the Sekem University will be the most lengthy and strenuous, but I feel I still have time for it.

Why establish a university like this in Egypt? Egypt with its seventy million inhabitants is among the largest of the twenty-two Arab countries. Because of its geographical location and its historical past it holds a position which enables special connections and relationships, although these are not usually even realized by the country itself. Egypt lies on the Mediterranean, making it a member of the Mediterranean countries, which makes it open towards Europe. It is also a part of Africa, and connected to Asia through the Sinai. Because of this, solely from a geographical point of view Egypt can be seen as having a bridging function.

From a religious point of view Egypt has had a connection to Christianity from the start. The first Christian monasteries were founded in Egypt, some of which still exist today. The evangelist Luke was an Egyptian physician. When Islam spread in the seventh century, ninety percent of the population became Muslim. Today, the Egyptian civilization influences all the countries of the Middle East through its media, especially its film productions and literature. In all Arab countries Egyptian teachers work in educational institutions. Egypt acts as an example for the Middle East, which has effects on different levels.

Entrance to the Sekem Academy.

From a more recent historical point of view Egypt was brought into closer connection to Europe through a renaissance in the nineteenth century. European researchers like Jean-François Champollion discovered ancient Egypt and for the first time Egyptians also realized the treasures of their own former great civilization. Napoleon's conquering expeditions took ancient artworks to European museums and the Europeans noticed that the roots of their culture were influenced by Egyptian art and science, which then became absorbed into the European pool of knowledge. The conscious awareness of this connection leads to a thankful remembrance of the origins of European spiritual life. On the other hand Egyptian statesmen like Mohammed Ali looked towards Europe to enrich Egypt with European spiritual life and profit from the resulting economic and legal abilities. Egypt became a wealthy country through this multicultural exchange, as I still experienced during my childhood.

Historically it can be seen how the connection between European and Arabian spiritual life was repeatedly fruitful for both countries. I am thinking for example of the thirteenth-century Emperor Frederick

II, who was brought up by Muslims during his childhood in Sicily, and who radically influenced the Christian-European culture. Friedrich II spoke Arabic fluently and knew the Koran well. Tragically the Pope of that time, Innocent III, expected him to continue the crusade against the Muslims in the Holy Land. But Frederick negotiated a peace treaty with the Sultan Al Malek who lived in Egypt. The memory of this European who came to the Orient not with a sword, but with an outstretched hand, still resonates positively today. The possibility for enrichment through mutual encounters and openness for the other is the reason why I do not want to start an Egyptian university without European input.

Nowadays relationships between Europe and Egypt are almost exclusively economic. A stable political situation is the most important condition for cooperation. But it is just as important to work together on a cultural level. Egypt's cultural development shows great deficits in this area. Economic progress and cultural advancement become detached from each other, creating the ground for conflict: religiousness turns into fanaticism, and people start rejecting anything foreign as they are not educated sufficiently to learn to take responsibility for their ideas.

I mentioned the special geographical location of Egypt as a bridge between Europe, Africa and the Arab world. But Egypt also had a bridging function in the cultural sense. This is where I see an important task for the Sekem university. Through cooperation with European institutions and universities, the graduates should become able to advance Egypt's development using practically orientated, innovative ideas. Sekem as a developmental initiative in Egypt has always strived towards these goals, and has successfully worked in close cooperation with Europe economically and culturally. The synergism of this constellation leads to the hope of a successful future for the Sekem university.

The Sekem University strives towards educating its students in many directions, using scientific freedom, because one cannot just continue learning in university like in school. Through participation in existing, applied research right from the start the students are encouraged to attain independent research ideas and individual solutions. This helps them to develop steadfast personalities with the ability to take on responsibility. The accompanying study of philosophical and artistic subjects helps to instil values and ideals which can then become the basis for social deeds. Finally I see the necessity to learn

Atos production meeting.
Overleaf: Sekem Head Office.

to act competently to be able to realize ideas practically. Once these goals have been attained, the influence of Sekem University with its new concept will be able to radiate throughout all the Arab countries because of Egypt's special position.

The artistic impulse

The artistic appearance of Sekem was of prime importance to me. I wanted beauty and grace not just in addition to the companies, but as an integral part from the start spreading its influence over everything. In this context there is a European historical event with which I am deeply impressed. The Emperor Charles IV built an entire castle, Karlstein near Prague, not for defence, not for economic reasons, but for the advancement of art, religion and science.

It needs a particular attentiveness to bring art into the desert. You can implement a lot in the desert, but you need sensitivity which is able to communicate with the desert's delicate being. The desert

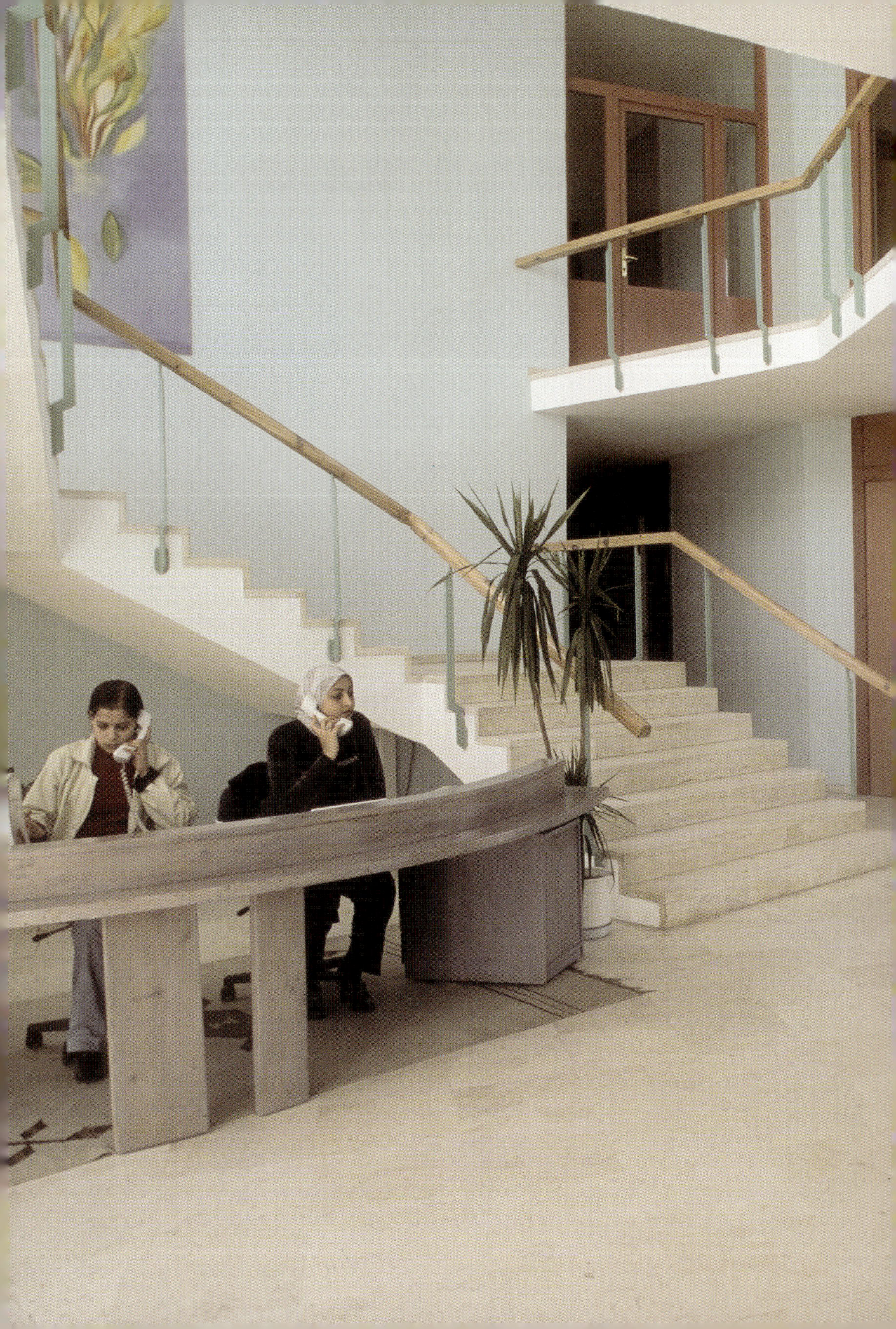

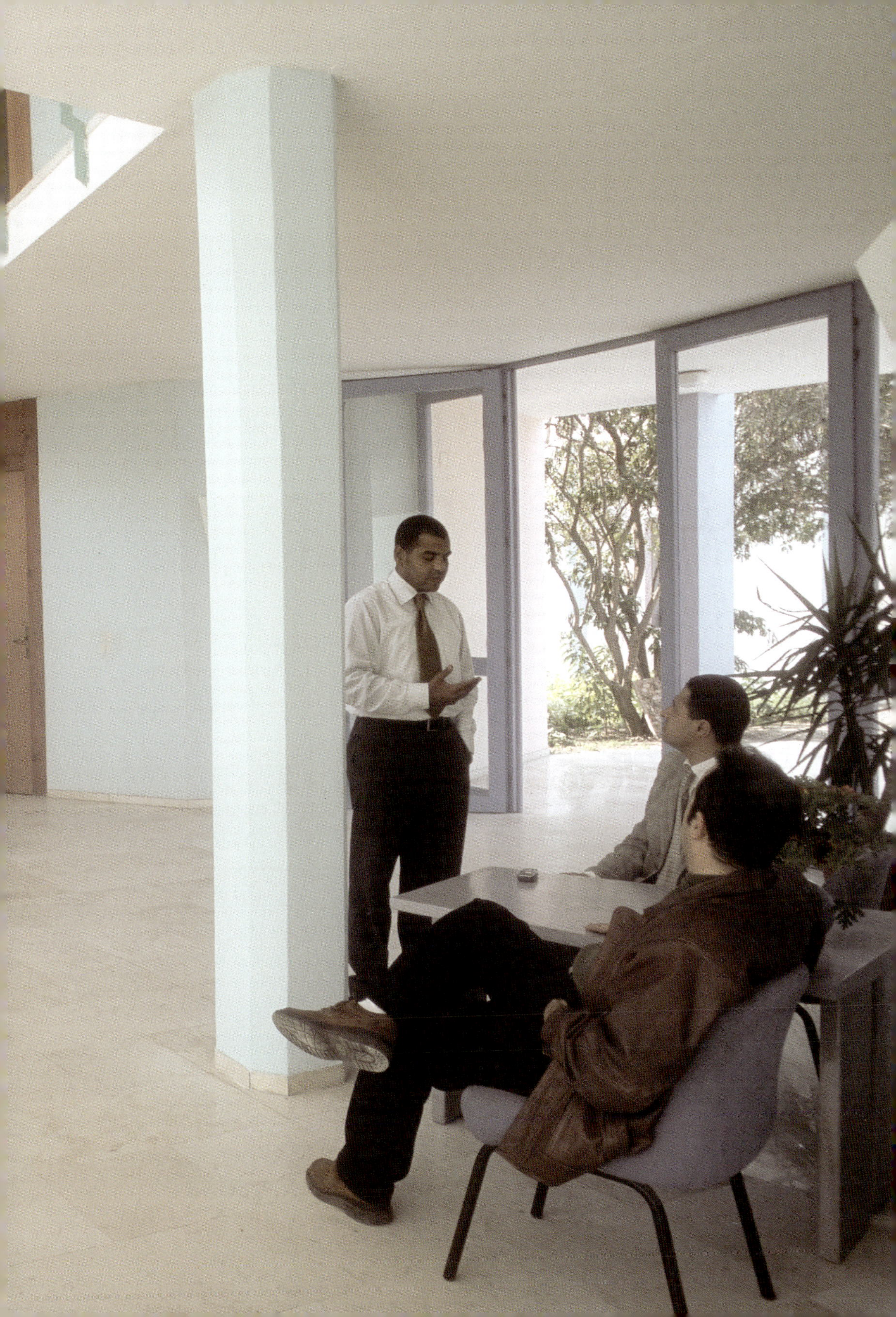

Amphitheatre with Hator building in the background.

should be approached with tenderness, because it has its own, very sensitive biological balance. I think of the recent irrigation attempts in the Egyptian desert as a brutal rape, which the desert will quickly reject if the people do not follow up their deeds with extreme consciousness and consistency. Many people experience the desert as a dead wasteland, because the opposing mineral and light-warmth elements clash together without the mediating factors of water and air. To be precise, I myself did not choose the middle of the desert to create a pure oasis, but rather placed my initiative in the border area between fertile land and lifeless desert. Animals and plants enrich the new land. With their help I was able to create a living foundation for advancing human development.

The teacher and writer Michael Bauer wrote the fairytale of the king's daughter Sinhold, which often occupied me during this period. In this story the princess sleeps in a palace for many years. Meanwhile a magician transforms her lands into a stony, barren desert. A king's

Work from the adult education courses, 1987.

son hears about her, and sets off to release her, but there is still the riddle that a man alone cannot free her, but that not more than one is permitted to. An old lady lives with her animals at the border of her country, but she dies shortly before the king's son reaches her, so that he cannot ask her for the riddle's solution. Because he finds the animals and plants without anybody to look after them, he decides to take them with him on his journey. With their help he is able to release the king's daughter.

The power of art gives hope and courage and through beauty works in a humanizing fashion. Art lives out of a strong centre; it needs a consciously created space to unfold itself while at the same time it occupies a border area. Art also leads to a sensory training, through which people advance themselves and liberate their senses.

'The smallest power in me is greater than the vastest sky!' What lives in Sekem because of this creative energy? For what has this space been made?

Sekem used to be a vast area of sand and stones which had to be transformed in an aesthetically pleasing way. Right from the start two aspects were of primary importance for shaping the grounds and later buildings: one the practical point of usefulness, the other the inclusion of artistic beauty. Fields planted with grains and medicinal plants and fruit plantations alternate with circular flower beds and avenues lined with decorative trees: Oleander and bougainvillea, large bushes and trees with wonderfully coloured blossoms, and in between palm trees with their beautiful fronds reaching up into the sky. The paths around the fields are right angled, while those around the living houses and social-cultural buildings follow more living impulses.

Underlying the concept of the grounds and building is a musical feeling: everything needs to harmonize. I was inspired by classical music during my stay in Austria. I longed for its sounds while setting up the venture in the desert. I have already described how I invited artists to Sekem, especially musicians, and how Osama Fathy organized the first grand piano for the Round House. I frequently asked visiting musicians, like Abel Weinfass from Cape Town, who decided

Sekem orchestra.

6. Education and Culture

to move to us many years ago, to play music before my many lectures. My idea was to make people listen intently, and through this establish a new breathing rhythm, so that it was easier for them to open themselves up to the more intellectual words and the message could reach the whole person. Nowadays Sekem has its own choir and orchestra. Concerts and large musical performances often happen at the Sekem festivals. In recent years the choir and orchestra has also worked with Arab music, and we are learning to hear what lies behind it and how to express it.

Looking after speech is also important. Through reciting the Koran this element already lives strongly in Arabian countries and is highly regarded. The classical Arabic language of the Koran and the Hadith lifts the Arab person out of his regional dialect and connects the different groups of people. Once or twice a year Dorothea Walter visits us, a speech teacher from Germany, and trains the teachers in articulation and voice expression. Although she herself could not speak a word of Arabic initially, she taught them to form the consonants and vowels much more expressively and to speak more fluently with rhythm and breathing. Using verses she researched the language and developed a practice booklet for shaping the Arabic language, which is used today by several Egyptian speech teachers. Our acting group, consisting of talented, enthusiastic company employees and led by the actor Yassar Badawy, practises plays for the big festivals which fill whole evenings and are received enthusiastically. It is always moving to see two thousand people sit in the circle of the amphitheatre beneath the swaying tarpaulin of the tent, and intently following the performance on the stage.

I feel the artistic movement of eurythmy needs to have a special position in Egypt. When I want to lead the Egyptian employees towards eurythmy I only need to show them ancient Egyptian pictures and reliefs. Using them as a background, I can explain that this art originates from Egypt, was taken up in Europe and transformed for our times. It was discovered that while listening to music and speech the larynx moves with the sounds and this activity can be translated into movement for the whole body. The person uses an ancient and the most natural instrument for this — his body — which becomes the expression of all he hears. The movements are not arbitrary, but deliberate, visible sound. Our eurythmy ensemble, consisting of Christoph Graf, Leonard Orta and Martina Dinkel, has been bringing eurythmy into all areas of Sekem for years: the kindergarten, the pedagogical and therapeutic institutions as well as exercises for doing at work. At the same time they work together on artistic projects, which they then perform publicly. We have Annemarie Ehrlich from Holland to thank for her valuable input towards developing a specific eurythmy for use in the work place.

Yvonne looks after the colour designs in Sekem, and has invited people from all over the world to Sekem to give painting courses to the company employees or in the school. I would like to personally mention Gerlinde Wendland from Germany, Suzanne Baumgartner from Switzerland and Annelie Franken from South Africa, who have supported us repeatedly with exhibitions in the Academy or school.

We have now established an art school in the vocational school building, where courses are held in all artistic areas. Artistic impulses also underlie the graphic design of our packaging and brochures. Because the packages represent Sekem to the outside world, and show the people who we are, it is particularly important to make sure they are artistically designed. It is not possible to protect things from being copied in Egypt, and sometimes I am happy to see something beautiful from Sekem has been copied for a different product. It often happens that our designs are imitated and the customers naturally sometimes mistake the products because the packaging is so similar. But we take this as an incentive to create something even more beautiful!

I see Sekem as a beautiful painting and the farm as the suitable frame. This painting is enlivened by the different colour nuances which the Sekem employees bring to it. Many visitors thank us in retrospect for the hope they have given through their visit, they thank us that Sekem exists. Then they pass it on and thus Sekem radiates out into the world. Through our artistic deeds a beautiful garden is created.

Celebrating the festivals

Islam has five main religious festivals a year and five times for praying a day. The dates of the festivals change, because they are calculated using the moon calendar, which changes by eleven days each year and takes thirty-three years to return to the same season. This means that the festivals move throughout the entire year. The prayer times follow the rhythm of the sun. Five times a day we Muslims pray at a time dependent on the position of the sun. So we experience a continuous harmony between sun and moon rhythms.

6. Education and Culture

Because I had the feeling that people had lost the awareness of the spiritual meaning of the festivals, I started to investigate them closely.

The month of Ramadan is a fasting month in Islam. For one month, believers are not supposed to consume any drink or food between sunrise and sunset. Nowadays the fasting month has become a month of celebration, and after sunset people eat more than in any other month. Towards the end, on the 27th day of Ramadan, the 'Night of Power' is celebrated. During this night we commemorate the Prophet Mohammed's first inspiration for the Koran — peace and prayers be upon him. As a forty-year-old he withdrew to meditate in a cave, and there the first verse of the first Sura of the Koran was revealed to him: 'Read the name which your Lord created.' (Usually translated in Sura 96) The wisdom of the Koran is revealed through the words of the Prophet. So this festival can also be described as a festival of wisdom. We tried to find new ways of celebrating the festivals in Sekem. Together we wrote and composed songs which impart the spiritual content of the festival. Now we sing these songs at the festivals in the schools and company buildings. Koran recitations and music, words about the religious background and a performance by the pupils round off the festival celebration and give it back its deep cultural value. The song for the 'Night of Power' is called 'The night of the occasion.'

The fasting month itself is also accompanied by a song, which encourages and summons strength for fasting.

A further Islamic festival is the ascension of the Prophet. The Prophet rode through the night on his white horse Buraq from Mecca to Jerusalem, from whence he ascended to the seven gateways of heaven. On his way he met all the prophets. In the highest heaven he saw an illuminated sycamore tree at the end of the road, and out of this experience he brought the five prayer times as a present from heaven to the people. I meditated for a long time over what this experience was trying to impart, und discovered that there is a connection between the growth processes of the plant and the times of prayer (see 'The sun-worshippers' p.148). The prophet could see the life rhythms of the plants revealed in the illuminated tree. The Muslim connects himself to the living growth processes through prayer. The song about the ascension of the Prophet Mohammed — prayer and peace be upon him — has a reverent mood full of praise and thanks and is called: 'Praise be to Allah.'

The Islamic prayer ritual encompasses a sequence of specific positions and movements, in which the person praying places himself

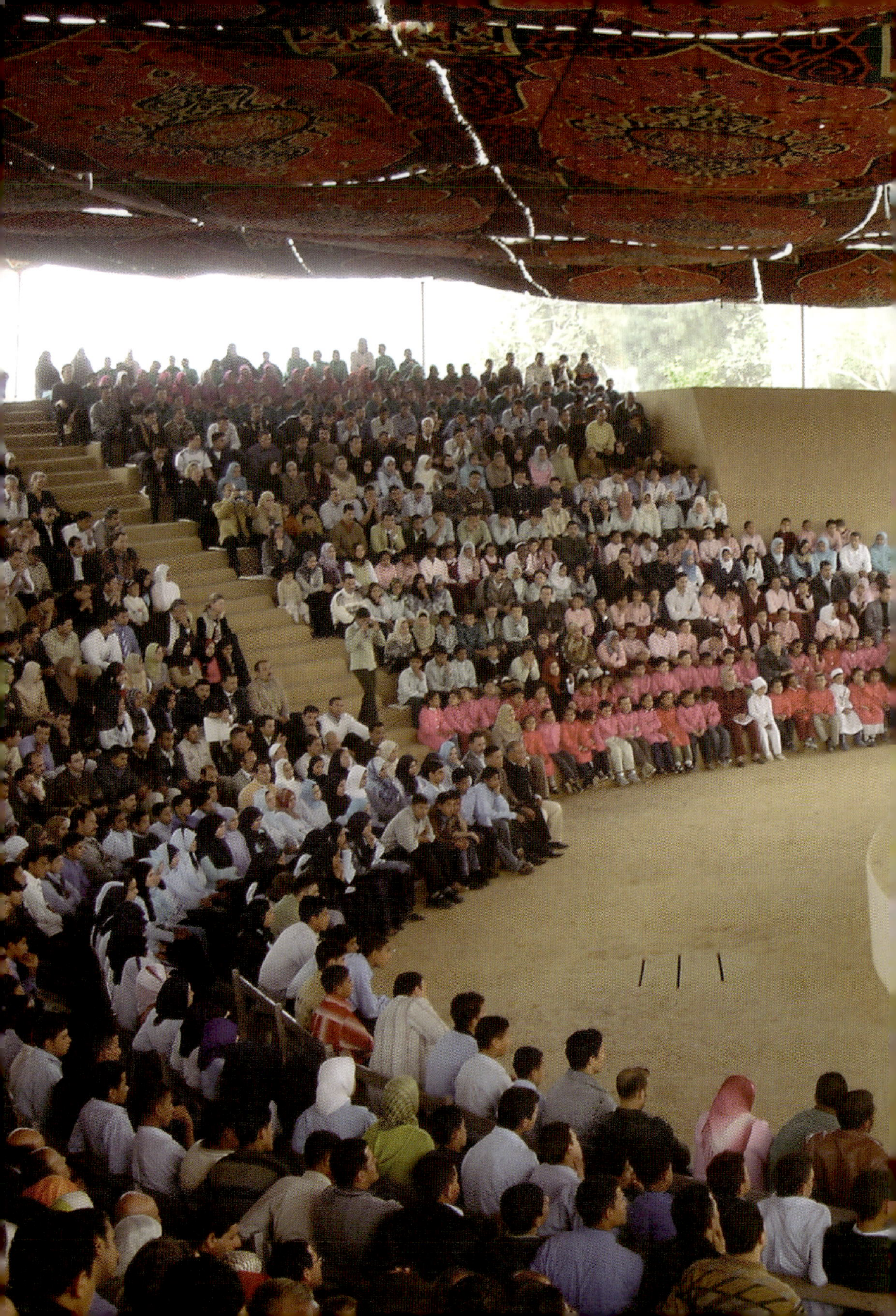

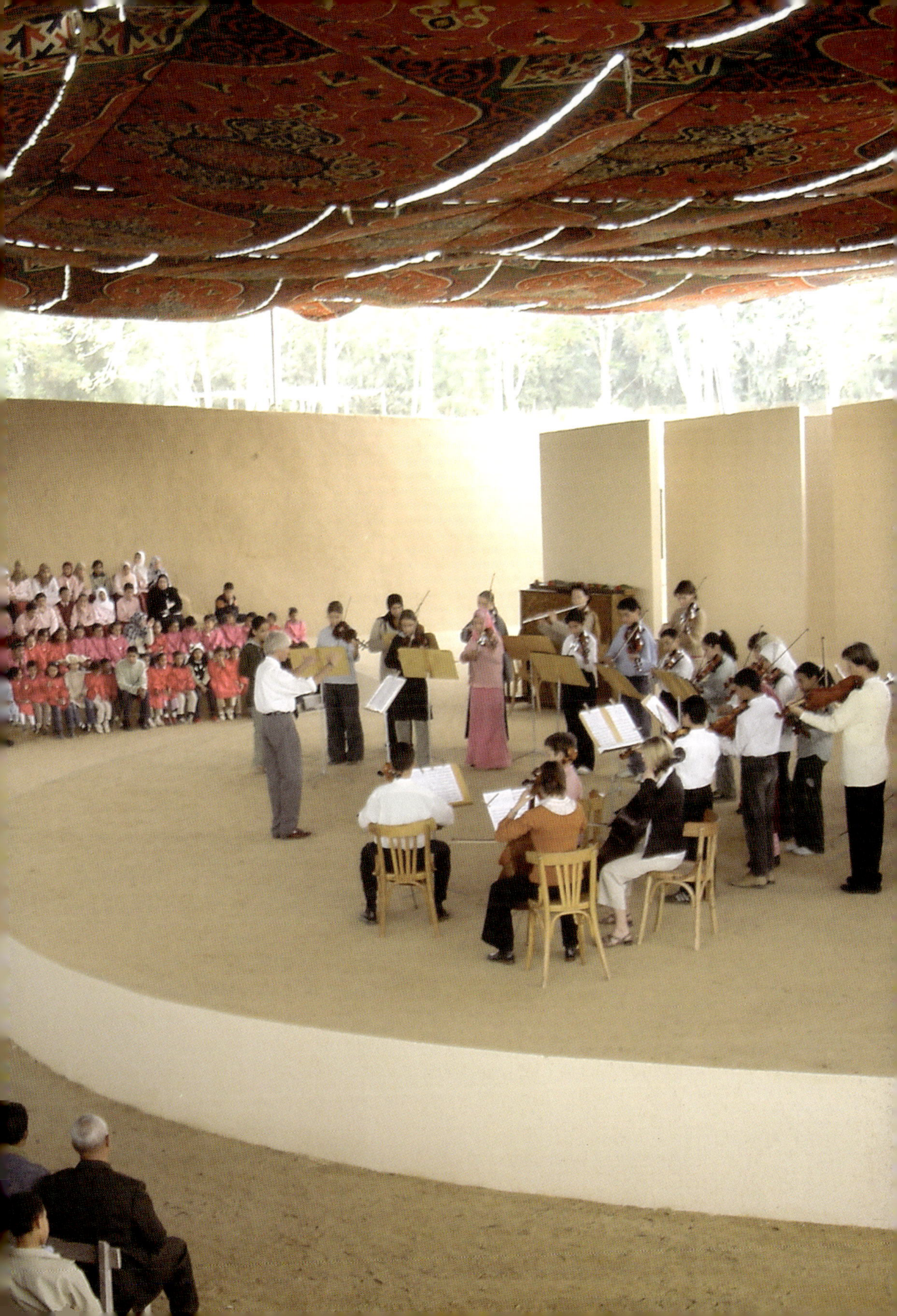

into a continuous relationship between heaven and earth. To start the prayer the hands are placed behind the ears in a slightly open angle — a listening gesture, opening up to above. The human uplifts himself to the divine by saying *Allahuakbar'* (Allah the most great) in this position. Then he places his hands onto the area of his solar plexus, the right hand crossing over and gripping the left hand. This position mirrors strong concentration and honourable energy, and points towards a very sensitive part of the body. The solar plexus, also called 'inner sun' is the centre of the body. From here it radiates in two directions: towards the chest and head region, which contains the light energy and is the seat of consciousness, and towards the metabolic-limb extremities, where the warm, chaotic natural energy is positioned. The human being is in harmony when these two polarities are combined in the solar plexus. In this position the Koran is recited as part of the prayer sequence. Then the hands touch the knees, the joint symbolizing a connection. The praying person then moves down to a horizontal position, bending right down to the earth to touch it with his forehead, hands and knees. During this process the solar plexus radiates downwards to the lower warmth area to connect to the powers rising from the earth. Bowing down to the earth is like a loving lowering into the earth.

This *raqaa* (sequence of prayer) is repeated two to four times during the entire prayer depending on the time of day. The prayer ends with the creed and the peace greeting to the angels on the left and right.

All Muslims in the world pray facing Mecca, so that during prayer times great circles of people praying surround the entire world.

After the ascension of the Prophet the direction of prayer was changed from Jerusalem to Mecca, to the Kaaba, the meteoric iron stone, which was erected by Abraham and Ishmael over four thousand years ago. The iron is a symbol of our willpower, and thus this festival has a will character, which is expressed in our song: 'Our will is as strong as iron ...'

The most important festival is the festival of sacrifice. It reminds us of the sacrifice which Abraham brought to his highest Lord. He is told to sacrifice his son Ishmael. The devil comes to Abraham and says: 'How do you know you received a revelation from Allah? It could also have been a dream.' Doubt creeps into Abraham's heart. But he instantly picks up seven stones from the earth and throws them at the

Previous page: Festival in the large theatre.   *Opposite and overleaf: Workers at Sekem.*

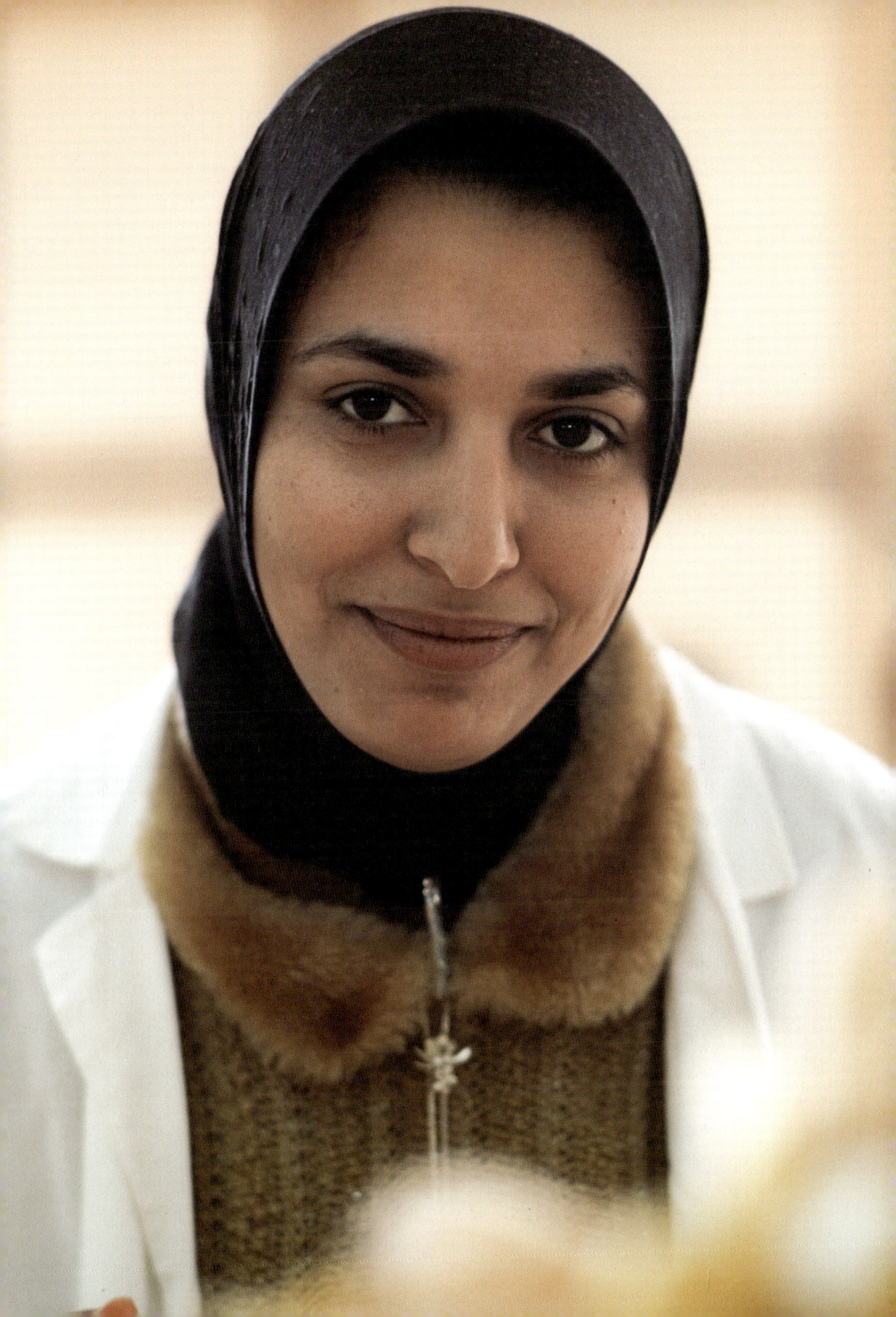

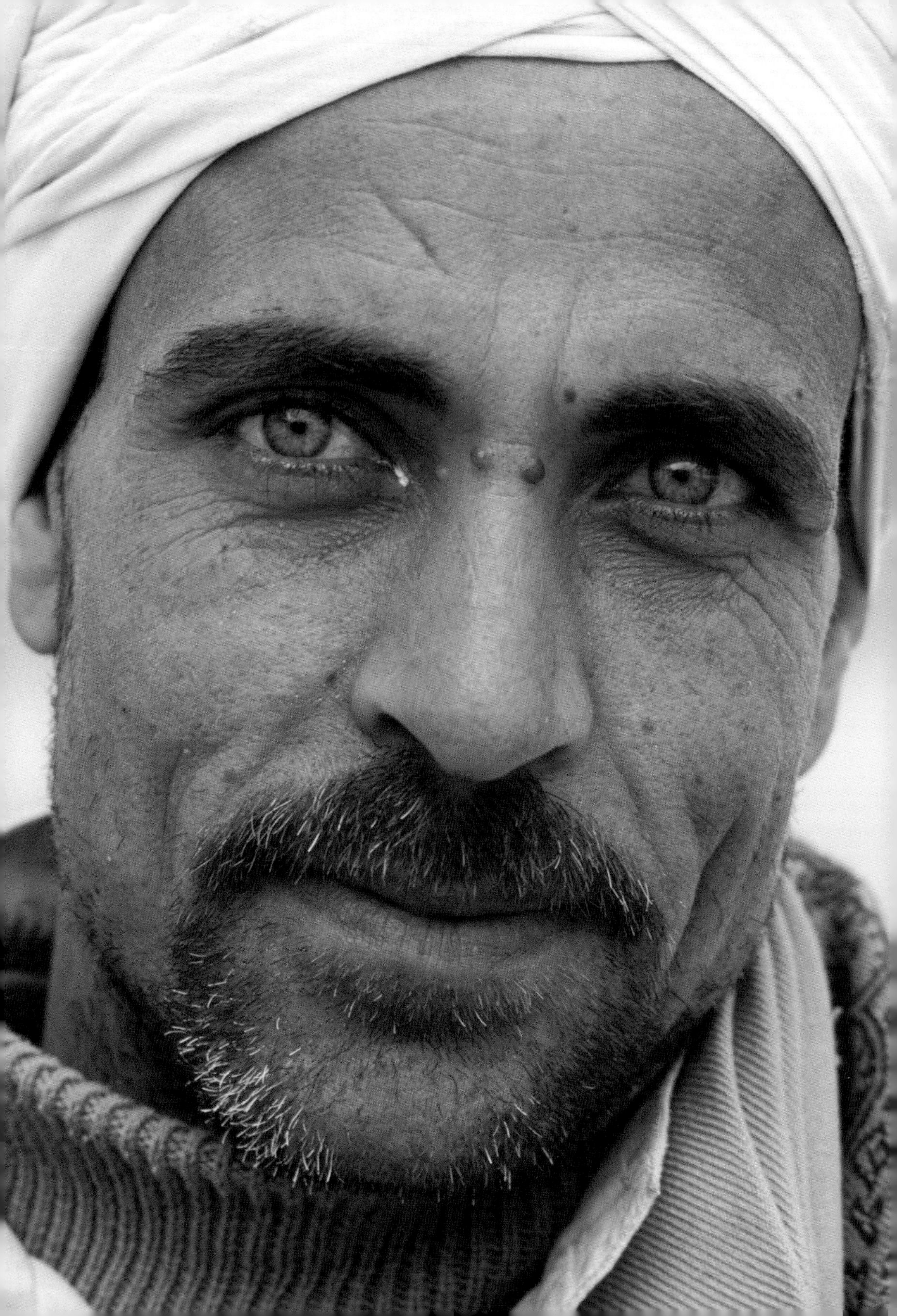

6. Education and Culture

devil. The devil also goes to Abraham's wife Hagar and to Ishmael and doubt also creeps into their souls. But the reaction is always the same: Resistance against temptation. The people who follow the correct leadership of Allah know how to protect themselves. In the Christian world this temptation is shown by Michael conquering the dragon. So in the festival of the sacrifice we celebrate this deed, this Michaelic deed, to use western terminology. Everyone able to make a pilgrimage to Mecca stands together in the exact same place where the deed described above happened more than four thousand years ago, takes stones and casts them at the devil. The song of the sacrifice festival is about the unwavering trust of Abraham in Allah: 'To you I am striving ...'

The Islamic calendar starts with the founding of the first Islamic community (Umma). A community should unite in the name of Allah. On July 16, 622 the prophet fled from Mecca to Medina (the hegira) and created a community which learns, works and looks after social contacts in the name of Allah. This is the Islamic New Year festival. An old, traditional song, which all Muslims know, tells about this event: *Talaa al bedru ...*

We also celebrate two further festivals in Sekem: one in autumn, the other in spring. The spring festival is on my birthday, not to celebrate me personally, but to represent the fact that every person has a day in which his individuality descended into earthly existence. The spring festival is thus a festival of individuality. All the children and employees and administration gather together in the large amphitheatre where music is played, a eurythmy performance or play is put on, and many speeches are given by the employees. The idea of the festival is to see oneself in the light of the community. It is an exercise for the people to take themselves seriously, to feel themselves as an individual part of the community and to strengthen the sense of themselves inside a community of others. In a cultural surrounding where the group spirit still plays an important part this is a vital awakening process.

In the autumn festival, where we celebrate Sekem's foundation, the companies and the school introduce themselves as a whole unit: all the employees go up onto the stage; then for example somebody recites a poem they wrote about what has been managed on the farm in Sekem. All the employees have achieved this success by working together; it is not the result of a single person. And yet the single person is still important. To illustrate this we remember special achievements, or celebrate anniversaries on this festival day. After working for seven years here, people receive a 'golden needle of honour' from Sekem.

Every time it is an uplifting experience for me and for the numerous guests when up to two thousand people are gathered together in the theatre.

Researching Islam

I have a deep friendship with Michael Heidenreich, a priest of The Christian Community. We became acquainted with each other during joint travels through Namibia and South Africa. His interest gave me the opportunity to talk about Islam, and we had repeated discussions about themes concerning Christianity and Islam. Together we planned a seminar week in Germany with the theme 'Christianity and Islam.' At the gathering I got to know the theologian Wilhelm Maas, who has worked extensively with Islam. Wilhelm Maas moved to Sekem for a year and we worked together closely for some time. Since then I have participated in many seminars in Europe which serve towards deepening and broadening the understanding of Islam.
In preparation for this task I asked my co-workers Martina Dinkel and Regina Hanel to help me with a German translation and interpretation of the Koran. We met regularly over a period of ten years and studied the Koran together. At the same time I also studied many other Koran translations. For our interpretation I penetrated into the spiritual aspects and we strove towards capturing the uniqueness of the Koran. On the whole I see my contribution towards Islam in a series of other efforts concentrated in an Islamic cultural setting. This gives me the impression that Islam is approaching a time of fundamental change. The Islamic world, and with it also Egypt, is lacking something comparable to what Martin Luther achieved for the Christian culture. The whole destiny of the modern world appears to point towards this task. For the vast majority in the world Islam is a mystery in its meaning and its striving. We hope this background becomes more apparent through our work.

7. Social Processes

New ways of working together

Legal aspects weave through all the economic areas and cultural institutions in Sekem, as all the different areas are in continuous contact with each other like a living organism. The cultural institutions could not exist without aspects from the economic sphere, and the economic companies could not exist without the educational ventures.

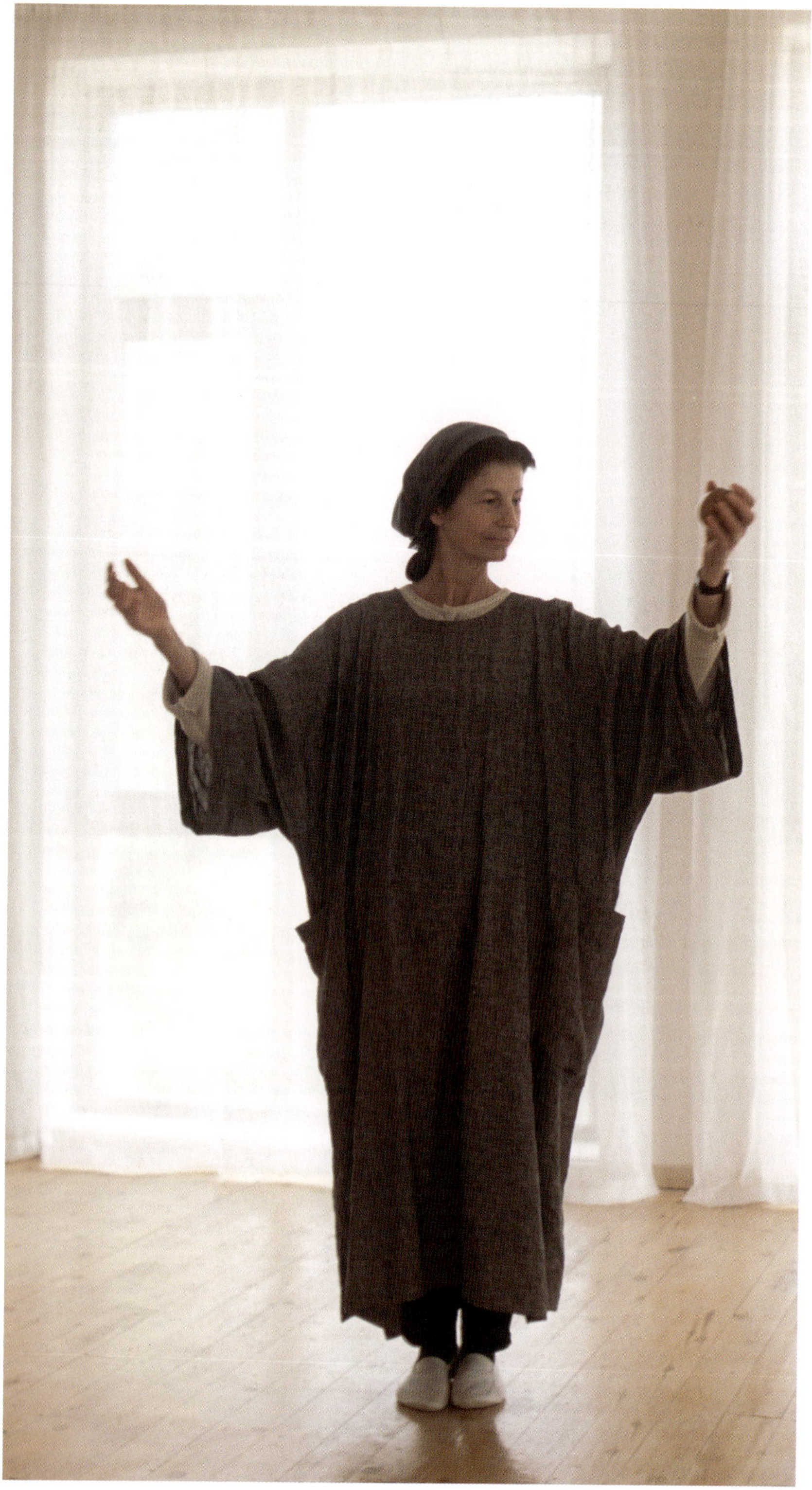

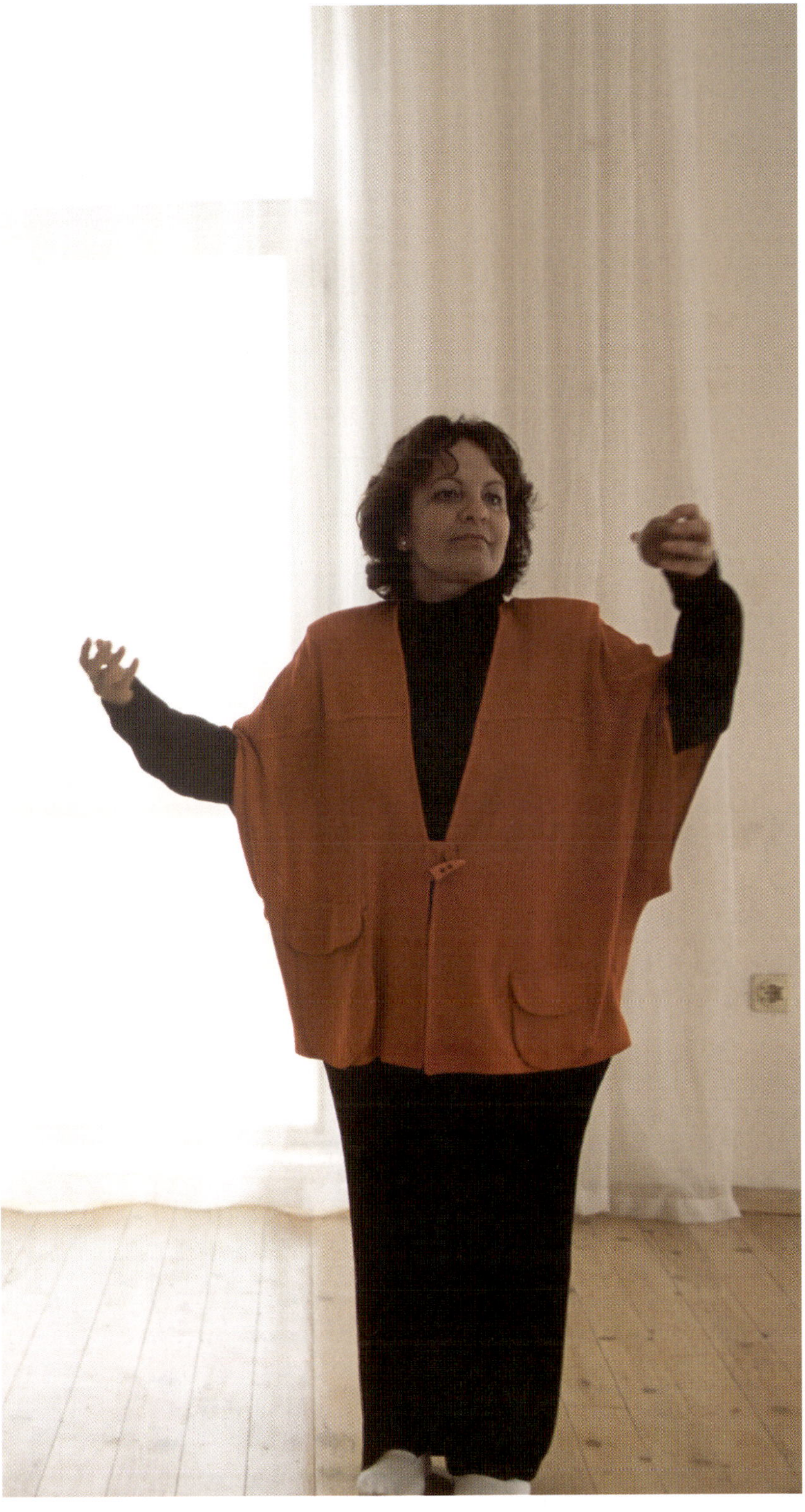

Previous page: Eurythmy.

All three areas are woven together, even though at the same time they are also independent bodies.

The circle is the characteristic shape for many gatherings in Sekem, from the daily start of work to the end of week assembly. In the mornings the employees of each company meet in a circle for a communal start. Each person briefly reports his intentions for the coming day. At the end of the week all the businesses and pedagogical institutions gather together. The circle is a symbol of social equality. Standing or sitting in a circle shows all participants are equal, an equalness which springs from the dignity of the human being. The individual person experiences himself as part of a community of individuals who are equals, and through this becomes aware of the other members of the community when he acts for the good of the whole. In so doing he takes on responsibility for his fellow people. At the same time every person as an equal has to represent himself to the others, that is, he has the right to be taken into the consciousness of his fellow people. Any agreements arranged as a result of this equality can only be changed using equality, for instance, by discussing the issues together.

Above and opposite: Morning circle with employees.

I gradually learnt that many people with whom I worked had a different relationship to time. This attitude prevents the ability to plan, set goals or analyse, and the people cannot correct or reflect on their actions. But the warmheartedness that I encountered made it easier to implement new ideas.

Several years ago we founded a cooperative for employees to organize the social processes, the Cooperative of Sekem Employees (CSE). It is an independent legal organ, which runs through all the businesses and cultural institutions. Setting up the CSE brought about a great renewal. The people around me had become accustomed to me being the authority for regulating and settling all social questions and arguments. But with the founding of the CSE other people gradually took over these tasks. They saw and picked up the chances enabled by the new legal forms and matured towards working more independently. Nowadays we have qualified social workers, psychologists and lawyers whose main task is to create and care for social forms. What are these forms?

While at the start of the venture I was relieved every time a person

Agricultural research meeting.

came to me in the desert and was willing to work for a few days or weeks, nowadays hundreds of people flock to our gates looking for work. The first department of the CSE is responsible for employing new members. Once the people have received affirmation of employment, then their rights and duties are discussed and they sign a contract, in which different aspects of social insurance are also taken into account.

But this is not all that the social workers have to do. They also

Managers of the Sekem group with the Abouleish family in front.

record further details of the employees, their family situation, state of health and education, their needs and requirements. On this basis they can form a picture of all Sekem employees. Taking previous education into account, they can recommend further education for the employees, for example English lessons for people who require a foreign language to perform their duties. If there are many new employees, then they make sure these people are informed about Sekem's basic ideas. Employees pay a small monthly contribution into a social fund which enables us to meet requirements and needs of the workers, for example everyone can expect additional financial support for a marriage, birth or death.

Some things are prohibited in Sekem (for security reasons), particularly for example smoking. Smoking is also damaging to health. We have had achieved good results through our health education events, which take place during the weekly employee meetings. Doctors, psychologists and drug advisors work together in this field.

The same department of the CSE inspects the working conditions of the Sekem employees. They visit the companies and talk to the workers, checking for possible faults or health hazards which could occur because of the specific working conditions, for example by breathing in dust. They then suggest improvements, for example new filter equipment. Some time ago this group suggested installing cold drinking water containers for all the employees.

A further task of the employee's cooperative is the improvement of the living conditions of the workers. Some of the bare necessities were lacking: for example, they ensured that the roads around Sekem were tarred, that telephones were installed and a post office built. One enormous problem was the hygiene and sanitary conditions of the workers and their families. Because of faulty infrastructure the people living around Sekem were forced to drink polluted water. With the support of the European Union a water treatment plant was built and all the houses received running water. Simple toilets were also built as part of this initiative. Gradually Sekem developed into a whole new village with its own village council and a police station.

The first department of the CSE is responsible for looking after the individual members of the community, the second department for social cooperation, which acts as a mediator when disagreements arise. When the businesses or cultural institutions in Sekem have a problem we regulate it by letting an independent employee of a different branch listen to the problem, ask questions and then reflect them back to the group. Experience has shown that this treatment alone is often sufficient to solve the problem.

This simple coaching procedure does not work on a managerial level, as the problems are often more complex. When they do arise in this area, a specially trained employee of the CSE comes to talk to the managers.

Several years ago we introduced a new practice called Key Performance Indicator to Sekem. This is an internal performance review of the employees, a process of evaluation where every employee is judged according to objective criteria: a social worker agrees on an individual goal with which the employee can identify himself, and his tasks are devised accordingly. Once a month the social worker and employee evaluate whether these goals have been achieved, and if not what could be possible reasons for the failure, whether the worker is in the right workplace or if it would be beneficial to transfer. There are also various training programs connected to these discussions. This procedure inspires the people to perform their tasks increasingly conscientiously. In the performance evaluation discussions the employees are also familiarized with the Koran verse in which Allah says: 'Work, because Allah, the Prophet and the believers can see your doing!' This leads the people to further advance themselves.

The CSE also conducts research. Professors of widely different subjects and politicians belong to this circle of researchers. Mahmoud Sherif, director of the Sekem academy, also takes part. This team of researchers is involved with basic legal questions and helps develop new social forms of working together though many discussions.

The Sekem fabric

Ever since the first people, both Egyptian and European, came to work on the farm, I wondered how to ensure that this growing community remained a living organism with a dynamic development. A beautiful saying of Goethe's expresses my highest ideals with regard to social forms of working together: 'And neither time nor power can destroy forms which have been shaped and have developed in a living way.' This quote reveals how the ideal community of people cannot be destroyed by any power in the world if it is willing to mature continuously and employ the necessary spiritual activity.

Since founding Sekem I have worked with questions relating to which circumstances give rise to creatively shaped and living forms, and how they remain dynamic.

The help of the spiritual world is required to achieve a living form, which is revealed to people who are open to inspiration through their own spiritual work. Because of this there has been continuous spiritual work among the Sekem employees from the start. This ongoing spiritual work radiates into all areas of our dealings and creates the solid foundation for the future development of Sekem.

I was able to understand the creatively shaped living form of a community of people concretely as a kind of 'life fabric.' During the first years I was responsible for the weave of the fabric. But over time, the interweaving threads became the tasks and responsibilities of many people, whose efforts all contributed and continue to contribute towards the success of the whole venture.

Right from the start I realized the importance of placing people at the connecting points of the network. Even if they were not able to perform their tasks initially, their deeds helped create a space which I was convinced they would grow into with practice, or that other people with the necessary abilities would come and fill this space. Creating such institutions is an important reality, as once established the venture can develop further with more ease.

This living human fabric was initially created by a single person. As it grew, it became more differentiated as the tasks were spread out to many people. And it will continue to grow. During the course of time the Sekem fabric has become richer and the weaving threads finer, as more and more people contribute to its creation.

The Sekem holding

In 2001, we decided to establish a holding entity to make the established enterprises more efficient and to use the synergy of the cooperation between the different businesses. This main task of this holding body is to administer the finances of all the companies, but it also oversees many other developmental projects. Inside Sekem we call this holding 'the developmental centre.' Continuous improvement and development are the signs of a flourishing business, which, as everyone knows, is difficult to achieve. So we have a department which deals solely with the task of helping each individual company with their developmental process. The economic term for this is 'business development.' This developmental centre is also responsible for the education and training programs for the employees. The specialist term for this is 'human resource development' (staff development).

Because our goal is to always perform to the highest quality standards in all our work and business areas, the developmental centre is also responsible for the Total Quality Management in all the companies. All our products and services are subjected to strict, internationally recognized quality guidelines, for example the Demeter guidelines, the Bio Swiss (ecological cultivation) guidelines, the US National Organic Program, different EU guidelines, ISO 9001 (quality management system), ISO 1400 (environment management system),

7. Social Processes

the Hazardous Critical Control Points, system for food security, the Good Manufacturing Practice, system for pharmaceutical companies, Euro GAP (Good Agricultural Practice, agricultural documentation system), and Fair Trade (inspection and certification system for fair trade). All these standards are initiated, followed up and permanently supervised by the holding.

We are convinced that an initiative like Sekem can only survive over a long period of time with the help of a local, regional, national and international network system. The different quality guidelines are important for achieving this goal.

The developmental centre is also responsible for public relations and communication with our partners, customers, government offices and other contacts, and coordinates all the activities in these areas. Modern information technology is an important tool to make internal and external communication processes more efficient and quicker. Because of this we are always looking for the best possible technology appropriate for our needs.

We strive to learn to work and work to learn, and through the activities described above we hope to be and remain a living organization — a never-ending process!

When I founded Sekem, I started as a single entrepreneur. In the economic world it is usual that what has been established by the entrepreneur is seen as his private property. He is also able to bequeath this property. This view seems neither adequate nor appropriate for Sekem. My conviction is that one can sell or bequeath a house one has built, but not a business with many hundreds of employees. For a long time I have been looking for other possibilities, also from a legal point of view, to find an answer to this question. In the long term we want to neutralize the capital. This would mean that everything belonging to Sekem is not private property, but put into the services of Sekem. For this purpose we are going to establish a foundation into which we can transfer all Sekem's productions.

Sekem's motto

Every year we hold a seminary as part of a few days of reflecting on the aims and goals of Sekem. These meetings have culminated in the formulation of a motto developed out of principles which we perceive as important and which we regularly update together.

> *We aim towards living together according to social forms which reflect human dignity and further development, striving towards higher ideals. Our main goal is a developmental impulse for people, society and the earth. Sekem wants to contribute to the comprehensive development of people, society and the earth, inspired by higher ideals. The cooperation of economic, social and cultural activities is stimulated by science, art and religion.*
>
> > *Sekem has set itself the following goals for the economic sphere:*
> > * *Healing the earth through biodynamic farming.*
> > * *Development and production of herbal remedies and any kind of product or service which relates to real consumer requirements and has standards of the highest quality.*
> > * *Marketing in associative cooperation between farmers, producers, traders and consumers.*
>
> *We aim to advance the individual development of the person through the cultural institutions. Education towards freedom is the goal of Sekem's educational institutions for children, adolescents and adults. Health care and therapies using natural healing remedies are provided by Sekem. The Sekem Academy for Applied Art and Sciences researches and teaches solutions to pertinent questions from all areas of life.*
>
> *Socially, Sekem furthers a community of people from all over the world who recognize the dignity of the individual, enabling both learning to work and working to learn, and providing equal rights to all.*

This motto is available to all in a printed form, but I do not believe that it would have a chance of being implemented without concrete, daily connection to the spiritual world. It is the attention towards the spiritual 'being of Sekem' which enables the awareness necessary for human encounters. Everyone coming to us wanting to learn is welcomed, and we try to integrate their abilities and talents. But they need to recognize what has grown here and be willing to take part in the communal developmental process. And because I know this creation is not mine alone, I am full of gratitude towards all those who together with me continue caring for this 'Sekem being.'

Outlook: the hill

I decided to keep one hill free of all buildings, and called it the Peace Hill. For many years it was a reminder of the desert landscape I had found here at the start. Originally I planned my grave on it, but once, when I was standing there alone, I thought: 'Such a special place is much too beautiful for a grave; many people should be able to be happy here!' We still did not have a place where all the members of Sekem could gather and meet together. And so I decided the amphitheatre should be built on this hill. The amphitheatre seats more than 1500 people and it is where we celebrate the many Sekem festivals, have theatre and eurythmy performances, and hold musical events.
'And where are you going to put your grave?' Elfriede Werner asked me once she saw the work for the theatre had begun. We walked around the farm together and looked for a new place, but could not find a suitable one. Simple Egyptians are not usually buried in a graveyard, another reason why I wanted to set an example of a dignified departure from this earth. One day, when I was inspecting the beehives, I noticed they were located in a beautiful, peaceful place. I thought the bees would be just as happy to move somewhere else, and started building a grave there according to the ancient Egyptian form: with a path to the west, built with golden yellow square stone blocks. We erected a chapel with a round roof in the north, where the person entrusted to Allah is carried after death. Verses from the Koran were chiselled into the yellow stones of the walls with themes concerning the reincarnation of human souls. These verses are: 'How can you deny Allah, as He gave you life when you were dead? Then He will cause you to die and restore you to life, and then you will be returned to Him.' (Sura 'The Cow,' 2.28). 'Have they not seen how Allah conceives of the creation from the beginning and then renews it? That is easy enough for Allah.' (Sura 'The Spider,' 29.19). 'Lo, We will bring back the dead to life. We record what deeds they have done and also the marks they have left behind. All things we have noted in a clear book.' (Sura 'Ya Sin,' 36.12). 'O you soul, you that have found

peace, return to your Lord contented and accompanied by His pleasure. Join the rows of My servants and enter into My paradise.' (Sura 'The Dawn,' 89.27–30).

On the eastern wall there is a verse that I composed:

> *When I die, oh Lord,*
> *I will return to you.*
> *I sowed the seed in Your name,*
> *And from You the harvest springs.*
> *I lit this candle,*
> *Oh Lord, preserve its light from the darkness of the world.*

My father had always taken an interest in the development of Sekem, visited us repeatedly and come to terms with the fact that his son was leading his own life. I could not share the ideals which lay behind establishing the venture with him, but I liked telling him about my optimism and joyful events. He never really gave up trying to convince me, out of fatherly concern, that something else would have been better for me, and kept suggesting alternatives to my Sekem duties. In the end he said many good things everywhere about Sekem and greatly valued me. Our relationship was characterized by great respect on both parts. I always addressed him formally and kissed his hand. He died at the end of 1982.

My mother also wholeheartedly followed Sekem's development. We visited each other often. After the celebration to inaugurate the school building she come up to me and said: 'I often pray for you, my son. I always pray: May all your wishes be fulfilled. Sekem seems like the prophet Abraham in the desert, who gathered so many people around him and spread blessings.' Towards the end she found the journey to Sekem very strenuous, although the roads leading there had literally become levelled, as she had always prayed for me: 'May all your paths be levelled.' I had to hold her mouth shut when she tried to say a further, too effusive prayer. Then she would laugh at me and we understood each other.

When my mother fell ill with cancer, I took her to the best doctors and we cared for her on the farm. Towards the end she had to lie in a hospital and I visited her daily. I often held her in my arms and recited verses out of the Koran for her. Throughout her life she had been a very devout woman. One day when I entered her room I felt a distinct change. I took her into my arms, held her face against my cheek and whispered loving words to her. I thanked her for her love, which had always accompanied me, and for her good deeds in the world. She died in my arms, quietly and in peace. I stayed with her for a long time and felt a warm inner joy: she was released and I had been allowed to accompany her birth into the spiritual world, just as she had physically given birth to me. Once I had returned to Sekem, the community celebrated a funeral for her where I talked about her life. Surprisingly enough, I did not feel empty after her death, but felt strength and energy coming from the other side.

Since the day we met a wonderful feeling of friendship existed between Elfriede, Hans and myself. We were a great trio. At least once a year they came to stay at Sekem for a longer period of time and we planned new projects and worked together spiritually. Elfriede founded the German Association for Cultural Development in Egypt, and led it with a lot of initiative. She organized journeys for us throughout Europe where I could talk about Sekem. We established a widespread group of friends who still accompany us with their thoughts today. Elfriede inspired people to come to Sekem, and encouraged them to give donations. The three of us went travelling to Yemen, to the mountains of the Sinai and to the oases of Egypt. We remained in constant contact. I will never forget how these two friends cared and looked after me during my period of illness. Eventually they had their own house on Sekem.

In the summer of 1999 Hans and Elfriede were staying in Sekem again. When the time for their departure came we said goodbye and they left. But a few hours later they suddenly appeared again: something had not worked with their flight. So we hugged again and talked until they left, this time for real. It was Elfriede's last visit to Sekem. Shortly after she fell ill with cancer. I visited her in Germany in August 1999 and together we looked back over the past eighteen years. Then Hans accompanied her lovingly on her difficult path. I came to see her once a month. She died in January 2000. After returning from the funeral I expected to feel a great emptiness in Sekem, but again this was not the case. Instead I experienced the energy for new tasks and a feeling of growing independence.

'Man cannot work alone'

I am often asked about the spiritual background of Sekem. Sekem developed out of my own vision. My spiritual inspiration came out of very different cultures: a synthesis between the Islamic world and European spirituality. I moved around freely in these different areas as if in a great garden, picking the fruits of the different trees. I would have felt I was restricting myself if I had had to limit myself to one way of thinking. But I felt there was enough inner space for everything in me. When writing about my student days in Graz I already described my experience of a syntheses between the Orient and Occident, the ability to rise above these two conflicting spiritual directions and become a third entity able to exist in both worlds. But I am also aware I am limited. After my death the forms that have been established in Sekem will have to continue developing in a living way. It needs people for this task who are able to guide Sekem according to the original vision, and who clearly know why Sekem was established.

A circle of people are entrusted with the actual running of Sekem, they constitute the centre of the venture. We call this group the 'council of the future.' One of the tasks of this council is to maintain a living connection to the well of spiritual inspiration. A further task is for the leading members to experience the connection to others as an enrichment and completion. Social ability also entails that every individual has an awareness of all the others, that he knows the conditions of the others and which tasks they are working on. Another task is the willingness to continue learning. A defining factor of a functioning shared leadership is that the people of the council of the future have more knowledge about the venture than the other employees. They know the background of decisions, they are aware of the risks and sometimes also of adversities which have to be met. The group is able to deal with these tasks with courage because of their trustworthy work together. During the gradual development of Sekem I always encountered questionable situations and great risks which I took upon myself because of my trust in Allah's leadership. But problems can be met with more objectivity if they are looked at from different angles. Through discussions with the outer world and the attitude that there is a solution for every problem people can grow and work together. They become able to stand up consciously for the development of the people and the world. Their dealings are led by the same trust that carried me alone at the beginning.

Today the Sekem initiative in Egypt is surrounded by the good thoughts of many friends, which are as strong and real as the earthly factors. We are very grateful for these thoughts. When I was studying in Graz in my younger days, I always felt how my mother was thinking of me, and how her love reached me there. Love gives trust in oneself and one's own abilities. Our many economic partners all over the world constitute another circle of people who accompany our work in Egypt, dealing with our products and promoting sales. I also want to mention all the people who support our artistic-cultural or research work. I experience it as culmination in my biography that I am able to recognize, respect, feel and look after all these different circles of people who give Sekem energy. The world is rich and great and my soul is vast. May my devotion be directed to Allah and his angels, without which nothing living materially can be created. The Koran states that Allah brings the people together, that we are not able to survive alone. I would like to place the consciousness of Him at the centre of my work, in the knowledge that everything we do and experience is not only of an earthly nature, but is supported by spiritual worlds. Allah says in the Koran: 'You, who can understand with your heart, can recognize the plan of god. If you recognize it and follow it you are supporting Allah. He will stand by you and strengthen your steps whenever you support him.' Is it not an overwhelming thought that the human being can support Allah? Because of this I appeal to humans to gather together and create real communities. One cannot work alone! That would be an illusion. Sekem has been created as something new out of both earthly and spiritual encounters. Because of this I think I can now say: the world would be a poorer place without Sekem.

Epilogue

At the time of finishing this story of my vision, new things are happening which appear to open a completely new chapter for Sekem.

Sekem was invited to a symposium of the renowned Council of Foreign Affairs in the USA to report about the activities of the Civil Society in an Islamic country. We reported about the history and the success of Sekem in front of important decision-makers in politics and public life. In the summer of 2003 we were selected by the Swiss Schwab foundation to be honoured for our work. Our business was seen as a blueprint for lasting social development by this distinguished Swiss foundation and we were invited to the famous World Economic Forum at Davos. Sekem sat at the same table as the most influential businesses of the world, and they let themselves be told about our activities. They wanted to know how we succeeded in implementing forms which were usually seen as ideal, but not applicable. And we explained that social, educational, and cultural elements were not just the 'icing on the cake' of our companies, but that all three areas, economic, cultural and social, were founded as a structured unity from the start with the necessary institutional backing.

Further international encounters followed on from this meeting. The World Bank asked Sekem to advise people how to create sustainable business ideas in a conference for young companies from fast-developing countries. Finally in September 2003 I received a completely unexpected telephone call from Stockholm: 'Congratulations — you have been selected to receive the Alternative Nobel Prize.' Sekem can now count itself among the many other honourable pioneers of social and humanitarian developmental work which have been honoured since 1980 by this foundation established by Jakob von Uexküll. In December 2003 we were ceremonially honoured by the Swedish parliament for services towards a better life.

This phenomenon of international recognition shows that Sekem is not only seen by the people directly involved — the children and staff of the farm — or only by the people of Egypt and our economic and scientific partners. Sekem is starting to have a place in a worldwide association of people and initiatives who are concerned with a healthier, more humane future on the earth. The 'net of life' created by Sekem and its initiatives is becoming connected to a larger, worldwide net. In this new phase our achievements are multiplied and perceived globally through international forums.

My vision now has a new, further level: to found a 'council of the future of the world' together with other institutions striving towards developing a better world. This 'council' would not be an abstract term, but carry a concrete message into the world: there is nothing more powerful than the invisible net of life, which connects people with their hearts. Its fabric is woven deeper than our understanding, and long before we first shake a hand we have moved along its threads. This net of life is more real than the most dangerous weapon, and unattainable for all outer violence. Only from it can real peace radiate. He who counts on its effectiveness is practising the most effective form of social art, because without using power or thoughts for advantage he can trust he will be carried by his energy and endurance. To learn to see the threads and to be able to form them determines the art of social working.

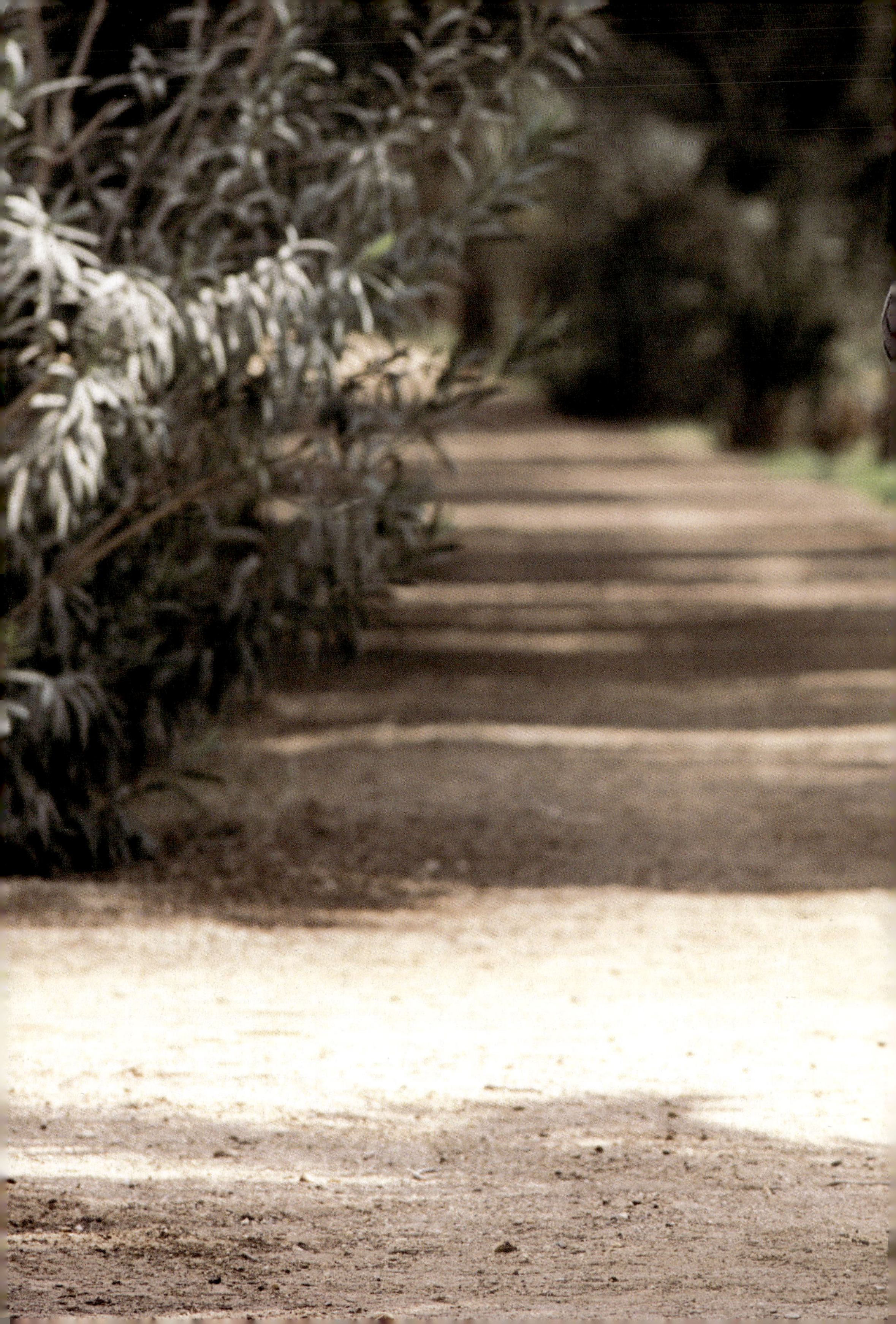

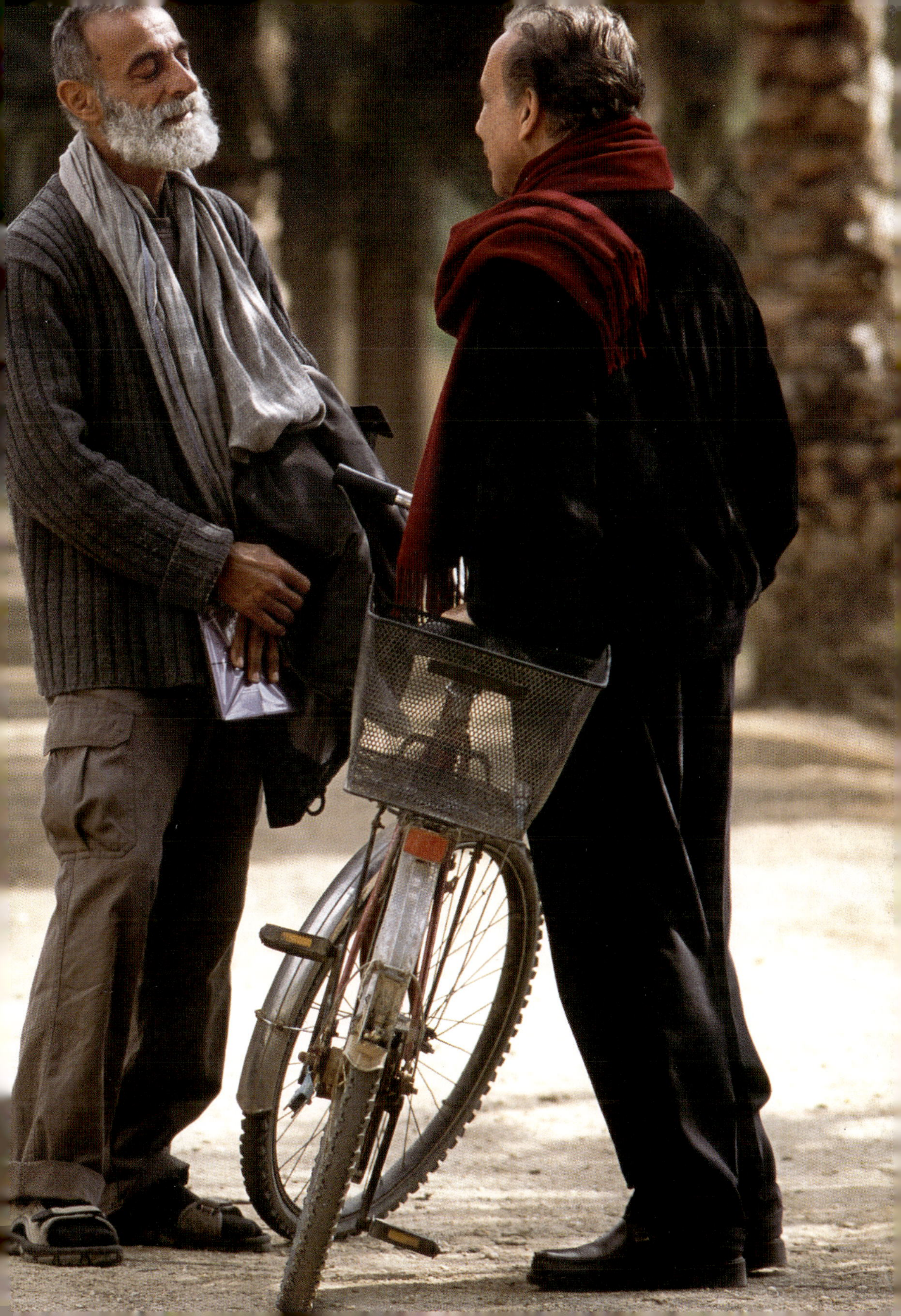

A Chronology

1937 March 23　Ibrahim Abouleish born in Mashtul
1956　Departure from Egypt to study in Graz, Austria
1960　Marriage to Gudrun Erdinger
1961　Birth of son Helmy
1963　Birth of daughter Mona
1972　Lecture in St Johann and encounter with Martha Werth
1975　Journey to Egypt with Martha Werth
1977　Sekem established on seventy hectares of desert northeast of Cairo. Start of biodynamic farming
1980　*Ammi majus* project commenced for the Elder Company in Ohio, USA
1981　Meeting with Elfriede and Hans Werner from Öschelbronn, Germany. This later led to the establishment of the Association for Cultural Development in Egypt
1983　Start of production of herbal tea remedies by Sekem
　　　Establishment of the Association for Cultural Development in Egypt
1984　Establishment of the Isis Company, start of marketing single herbal teas
　　　Establishment of the Egyptian Society for Cultural Development
　　　Export of Sekem herbs in association with the German Lebensbaum Company
1986　Establishment of the phyto-pharmaceutical company Atos as a joint venture between the German Developmental Society, Schaette Ltd and Sekem Ltd
1987　SCD establishes the Mahad Adult Education Training Institute
1988　Establishment of the Libra Company
　　　Start of the Sekem kindergarten
1989　Inauguration of the Sekem School
1990　Sekem pioneers cultivation of organic cotton
　　　Establishment of the Centre of Organic Agriculture in Egypt (COAE) for inspecting and certifying organic products.

1997	Establishment of the Conytex company
	Establishment of the Egyptian Biodynamic Association (EBDA), to provide agricultural training
1995	Start of clinical trials between Atos Ltd in specific universities
1996	Establishment of the Hator Company
	Start of the Medical Centre
	Establishment of the International Association of Partnership between Sekem and European partners
1997	Establishment of the Vocational Training Centre
	Establishment of the art school
	Isis starts selling in Nature's Best Shops in Maadi, Heliopolis and Zamalek
	Sekem, Atos, Conytex and Hator receive the ISO 9001 certificate
1998	Sale of the Viscum preparation through Atos
2000	Establishment of the Sekem Academy for Applied Arts and Science
2003	Ibrahim Abouleish was selected as an Outstanding Social Entrepreneur by the Swiss Schwab Foundation
	Award of the Alternative Nobel Prize to the Sekem community and Ibrahim Abouleish

Glossary

Ammon, Egyptian God
Bairam, festival at the end of Ramadan
Bedouin, nomadic desert people
Bio Swiss, organic cultivation guidelines
CSE, Cooperative of Sekem Employees
DEG, German Development Society
EBDA, Egyptian Biodynamic Association
Euro GAP, good agricultural practice and documentation system
fanouz, small lanterns
fetta, round, flat loaves of bread
fez, also called tarboosh in Egypt, a red tall Turkish hat with a twisted tassel
galabias, wide, woollen garments
GMP, Good Manufacturing Practice, system for pharmaceutical company
Gomaa, chief of a Bedouin tribe
GTZ, German Society for Technical Cooperation
HACCP, Hazardous critical control points, food security
Hadith, records of Prophet Mohammed's speeches
halal, permitted food for Muslims
halva, sweet made of honey and almonds
Hator, company with fresh vegetable trade
imam, prayer leader in the mosque
ISO 1400, environmental management system
ISO 9001, quality management system
Kaaba, meteoric stone in Mecca
Menja, city of Akhenaton
norak, a board used for threshing
rababa, musical instrument with two strings
Ramadan, the month of fasting
raqaa, sequence of Muslim prayer
SCD, Egyptian Society for Cultural Development
tajeb, exquisite food
Tomex, garlic preparation, first medicine of the company Atos
Viscum, mistletoe preparation of the Atos company
zakat, giving alms

Photographic acknowledgments

Markus Kirchgessner: Pages 2, 61, 97, 102f, 110 (both), 111 (both), 115, 118f, 121, 122, 123 (both), 126f, 142f, 153, 157, 161, 163, 164, 172, 174, 175, 178, 179, 181, 182 (both), 187, 188f, 192, 199, 200, 202, 204, 205, 207, 222.

All other photographs were supplied by Dr Abouleish or Sekem..

Index

Numbers in italics refer to photographs

Abouleish, Gudrun 39–41, *41, 168*
Abouleish, Helmy 41, 43f, *44,*
  90–92, 134, *138,* 139
Abouleish, Ibrahim's father 19,
  21, 23, 33f, 216
Abouleish, Ibrahim's maternal
  grandfather *18,* 18f
Abouleish, Ibrahim's maternal
  grandmother *18,* 27
Abouleish, Ibrahim's mother *21,*
  26f, 34, 216f
Abouleish, Kausar, Ibrahim's sister
  22f
Abouleish, Konstanze, Helmy's
  wife 60f, 139f
Abouleish, Mohammed, Ibrahim's
  uncle 30
Abouleish, Mona, daughter 41,
  43f, *44*
Abuchatwa, Chaled 130, 136
Afifi, Youssef, entomologist 138
Albrecht, Martin 106
Ali, Mohammed 63
Allah, ninety-nine names of 38
Alternative Nobel Prize 11, 87,
  108, 220
Ammi majus extract 87–91
Arabic language 195
Araby, Dr El 137f
Aswan dam 58
Atos pharmaceuticals *154,* 155–58
Azhar University, Al 59

Badawy, Yasser 193
Bandal, Tobias 155
Baumgarnter, Suzanne 194

Bedouin 74, 80–84
Beltagy, Dr El 140
Bentele, Gerdi 173
Boecker, Christina 155

Cairo, Sekem office in 93f
Champollion, Jean-François 185
Charisius, Klaus 181
Charles IV, Emperor 187
Conytex, company for organic
  cotton 140–44, *181*
Costantini, Margret 113
Cotton, organic 136–44

Dinkel, Martina 193, 202

Egypt, position and task 184f
Egyptian Biodynamic Association
  134
Ehrlich, Annemarie *165,* 194
Engelsmann, Volkert 125, 129
Erdinger, Gretel 39
Erdinger, Gudrun, *see* Abouleish,
  Gudrun
Erdinger, wife of Kajetan 39f
Erdinger, Kajetan 39f
Euroherb Co 125
eurythmy 193

Faruk, King 28
Fathy, Hassan, accountant 87
Fathy, Osama, musician 98, 192
festivals 194–201
Fintelmann, Professor Klaus 114,
  173, *177*
Floride, Christophe 154

Floride, Yvonne 154
Frank, Dr Roland 160
Franken, Annelie 194
Frederick II, Emperor 186

Gaballah, Mohammed 136
Gamasy, Chairy El 99f
Ghalib, Professor Haider 156
Goethe, Johann Wolfgang von 30f, 211
Gögler, Frieda 106
Graf, Christoph 193
Graz 34–37, 46f

Hanel, Regina 168, *170,* 202
Hartung, Thomas 125
Hashim, Yusri 135
Hator, company *126f,* 129f
Hedi, Abdel 163f
Heidenreich, Michael 202
Hess, Heinz 125
Hofmann, Angela 106, *107*
Hofmann, Georg 164

Innocent III, Pope 186
irrigation 79f
Isis tea products 153f
Islam 194–201
— and Sekem 146–49
Israel 48f, 51

Karnak 56f
Kindergarten *166*
Kläger, Eberhard 181
Kohl, Helmut, German Chancellor 180
Koolhoven, Bart 125
Kuschfeld, Monika 105

Laue, Erika von 113
Lebensbaum Company 124f
Libra Company 128, 130–36
Luxor 56

Maas, Wilhelm 202
Mahad, adult education 167f
Mahfouz, Dr Mahmoud 157
Malek, Sultan Al 186
Marienfeld, Dieter 162

Marienfeld, Ingeborg 162
medical center 157, 159f, *160f*
Merckens, Georg 61, 61f, 101, 124, 129, 131f, 134
Merckens, Johanna 135, 141
Merckens, Klaus 135
Mubarak, Mohammed Hosni, President 49, 180
music 192

Nasser, Abdel, President 28, 45, 58

obelisk 56
Orta, Leonard 193
Öschelbronn 113f

Pharmaceutical production 150–58
Raileanu, Maria 141
raqaa (prayer sequence) 198
Rashad, Ahmed 131, 134
Rehn, Götz 125
Reindl, Winfried 107, 125, 155, 167, 173
Round House 85f, *86*

Sabet, Berta 20
Sadat, Mohammed Anwar el, Egyptian President 48f, 101
Saher, Abdel, Dr 137f
St Johann 51f
Schaette, Roland 62f, 107, 124, 155
Scheffler, Dr Arnim 157f
Segger, Peter 125
Sekem, beginnings 74
Shabaan, Aiman 152
Shabaan, minister 101
Shauky, Ahmed 87
Sherif, Mahmoud 211
Siku, General Ali 101, 104
speech, as art form 193
Sorour, Dr Ahmed Fathy, leader of Egyptian Parliament *110*
Spielberger, Hans 135
Spitz, professor at Graz medical faculty 46
Steiner, Rudolf 52–54
sun worshipping 144ff

Index

Takis, Mr, from Cyprus 128

Ulrich, Wilfried 181
University at Sekem 184–86

Valley of the Kings 56, 105
Vocational Training Centre 180f, *182*

Walter, Dorothea 193
Walter, Ulrich 124f

Water, wells 77–79
Weinfass, Abel 192
Wendland, Gerlinde 194
Werner, Elfriede 105f, 113f, 173, 215, 217
Werner, Hans 106, 113f, 157–59, 217
Werth, Martha 52–56, *55,* 61, 64f

Zahran, Kamel 71–73
Zwieauer, Dr Johannes 65

Ecovillages: A Practical Guide to Sustainable Communities

by Jan Martin Bang

Jan Martin Bang explores the background and history of the Ecovillages movement, which includes communities such as Kibbutz, Camphill and Permaculture. He then provides a comprehensive manual for planning, establishing and maintaining a sustainable community. Issues discussed include leadership and conflict management, house design, building techniques, farming and food production, water and sewage, energy sources and alternative economics. The final chapter brings it all together in a step-by-step guide. Includes over twenty 'Living Example' case-studies of communities from around the world.

www.florisbooks.co.uk

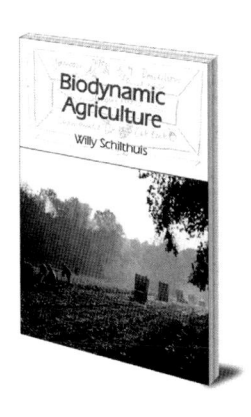

Biodynamic Agriculture

by Willy Schilthuis

In biodynamics the farmer or gardener works with the spiritual dimension of the earth's environment, and strives to maintain a sound ecological balance.

This colour-illustrated concise introduction to all aspects of biodynamic agriculture demonstrates that biodynamic crops put down deeper roots, show strong resistance to disease and have better keeping qualities than conventionally-produced crops.

www.florisbooks.co.uk

Waldorf Education

by Christopher Clouder and Martyn Rawson

This book is a concise introduction to the Steiner-Waldorf school and its philosophy. The two authors, both experienced teachers, describe Rudolf Steiner's innovative ideas on children's mental, physical and emotional development. The book includes examples from the classroom and the curriculum, to make this an informative guide for teachers and parents. Fully illustrated in colour.

www.florisbooks.co.uk